RAISING YOUR OWN MEAT

For Pennies A Day

Will Graves

Garden Way Publishing Charlotte, Vermont 05445

To all those long-suffering calves, sheep, pigs, rabbits, turkeys, geese, ducks and chickens, who produced so magnificently, in spite of my early mistakes.

To Ernest and Helen Graves, for their encouragement.

To Richard M (Dick) Ketchum, who, as editor of Blair and Ketchum's Country Journal, has guided and assisted nearly a generation of country writers.

And to my Garden Way editor, Fred Stetson, for his patience, suggestions and unflagging enthusiasm.

Photographs on pages 48, 117, and 129 by Frederick W. Stetson. All other photographs by Grant Heilman Photography.

Illustrations by Elayne Sears

Copyright 1983 by Garden Way, Inc.

Printed in the United States
First printing March 1983

Library of Congress Cataloging in Publication Data

Graves, Will.
  Raising your own meat for pennies a day.

  Bibliography: p.
  Includes index.
  1. Livestock. 2. Poultry. 3. Rabbits. I. Title.
SF65.2.G73 1983   636.08′83   83-1595
ISBN 0-88266-330-5

CONTENTS

INTRODUCTION

Several years ago, my life reached a turning point, and I made an important decision. I gave up a good, conventional job, sold a comfortable home in Middle America's suburbia, bought a 160-acre farm in Nova Scotia and moved north with my family. Together, we raised chickens, turkeys, ducks, geese, rabbits, pigs, sheep, beef cattle and other barnyard animals. On the new homestead, experience gained earlier on a family farm served me well. I was not a beginner, but I still made errors. Fortunately, my animals and birds produced magnificently, despite my mistakes.

The Canadian winters were severe, and, on many mornings, temperatures dipped to 20 degrees below zero. Bitter winds blasted down from the northwest, bringing heavy snows. On those days, I would have preferred to have sent the hired man out to do the barn chores. But, there wasn't any hired man, so I had to send myself. Sometimes, I'd shovel a path through the snow to the barn, then I'd break the ice in the water troughs and throw hay down to the animals. Those chores got my blood circulating—fast. And, the sight of those healthy, handsome and animated birds and beasts always warmed me from inside.

I enjoyed the animals, and I think that's very important. Enjoy them, but don't make pets of them. If you have never raised poultry or small livestock before, this book is for you. It's not for commercial producers. I provide basic information for those who want to provide their own meat for their own table. A supermarket may sell some pre-packaged meats at a cost lower than you can produce them yourself. But, the meat will never taste as good as yours. And, furthermore, through animal husbandry, you'll be restoring a personal touch to the production of your food, a personal touch that can't possibly be duplicated over the gleaming counters of a supermarket.

I'm making some assumptions in this book. First, as indicated already, I'm assuming you have little or no experience. I'm also assuming you don't have the necessary equipment or know-how to grow your own feed grains; therefore, you'll have to buy most of your own feed from suppliers or hardware stores. I'm also writing this book, at the suggestion of Roger Griffith of Garden Way, so it's specifically aimed for the person who wants to produce meat on a small scale. There are many ways to do this, but Roger and I agreed each chapter would present a simple, step-by-step approach, a definite system, if you will, showing you *a particular way* to raise a particular bird or animal.

I'm going to be discussing broiler chickens, turkeys, ducks, geese, rabbits, sheep, pigs and beef calves. In each chapter, my system or approach covers these points: when, where and what to buy; necessary equipment; approximate start-up costs; the time involved; and low-cost housing, pens and fencing. Also, there are sections on health, sanitation, daily care, observation and slaughtering. And, if you follow my system, you'll be able to calculate how much edible meat you can reasonably expect to gain. Prices quoted in this book are offered only as a guide and are obviously subject to change. In an attempt to make each chapter complete, some unavoidable repetition has occurred, but I've tried to keep it to a minimum. If you would like a quick comparison of the costs and time involved in raising the eight different birds and animals discussed in the book, see Appendix A.

Tending, caring and feeding poultry and livestock can help develop a sense of responsibility in your family, and you can't buy that experience at the store. You will discover things about yourself, and perhaps draw upon inner resources you'd always hoped for, but never knew you had. You'll find out if you really like working with animals. And, if you don't have it already, you'll learn patience. You've got to have patience. You can't push animals any faster than they really want to go, or are capable of going.

As a beginner, you might start with chickens. Relatively speaking, start-up costs are low, and, if you decide you don't enjoy raising broilers, you can get out of the chicken business in a hurry. Eight weeks and they're gone—unlike a lamb or pig, which take five months to raise, or a calf, which can take a year or more.

I acquired my early animal-raising experience in Illinois. While my childhood companions were away at camp or on the seashore, I spent my summer vacations on my grandfather's farm. In those days, every farm had chickens, ducks, pigs, lambs and calves, and they were raised to supply meat for the table. Farms were not as specialized as they sometimes are today.

For me, the hardest part of animal raising is slaughtering time. No one else in the family had the stomach for this job, so it fell on me, and rightly so. However, I did notice that no one ever had to be called twice for dinner, when there was roast goose, smoked ham, lamb chops or sirloin steak on the table.

After military service during the Second World War, I studied pre-veterinary medicine at the University of Illinois. But, after two years, I decided that becoming a veterinarian wasn't my goal, and I switched to other fields of endeavor. By the late 1960s, I found myself caught up in the "back to the land movement", and I finally decided to do what I wanted to do all along: buy a small farm and raise animals. That's the decision that took me to Canada, where I enrolled in sheep production courses at the provincial college of agriculture. A few years ago, I sold the Nova Scotia farm, and moved back to the United States, where I managed a couple of beef cattle operations in upstate New York.

Over the years, I've noticed the increasing efficiency of farm meat production. Before the War, it took five pounds of feed to put one pound of gain on a meat-type chicken. Today, it only takes two pounds of feed to produce the same results. Of course, the earlier feeds were often home grown and lacked some of the nutritious elements of the highly-concentrated, complete mixtures manuactured by milling companies today.

We have gained in efficiency in food production, but our heavy dependence on non-renewable resources to provide energy needed for modern efficiency is something that must be ultimately reckoned with. We saw the crippling effects of the energy crisis of the 1970s. The 1980s may bring troubles for our food supplies. Energy shortages, trucking strikes and consolidation of food production in large corporations are all developments that could hamper the delivery of food to consumers. And, at the same time, these developments make it more and more imperative that each of us learn to raise our own poultry and livestock for meat.

W.G.

CHAPTER 1

BROILER CHICKENS

If you are considering raising chickens for meat, here is a method for producing broilers and fryers. If you follow this system, you can net about 75 pounds of poultry meat within eight weeks, enough for at least 20 Sunday dinners for a family of four.

This method is based on purchasing 25 day-old chicks of a quick-growing, meat-type breed, feeding them commercial chicken feed for about eight weeks, and then butchering and dressing them out yourself. You should own or have access to a good-sized freezer.

The 25 chicks plus postal delivery charges (if you don't live reasonably close to a hatchery) will cost about $20. In order for them to gain a live weight of 4 pounds within eight weeks, you will have to feed the flock a total of 200 pounds of a complete, high-protein, chicken ration, which will cost about $22. Necessary equipment such as waterers, feed troughs and a brooder lamp with guard and reflector will cost about $28. Assuming that you don't have to build a structure to house the chicks, but already have some sort of building such as a shed, barn, garage or utility house suitable for raising a small flock, your start up costs will be about $70. Remember, the equipment can be used over and over again, if you decide to continue raising your own chickens for meat.

You will devote about 15 minutes twice a day, every day, to feeding, watering and observing your chickens; or about three and one-half hours a week. As with all penned creatures, the baby chicks must be cared for every day, seven days a week, no Sundays or holidays off. However, with a bit of supervision, a responsible child of 12 or more can tend the chickens. It is a responsibility. You are responsible for 25 living, breathing, defenseless, little birds. Their very lives are in your hands.

But, when you become personally involved in this kind of project, providing the meat for your dinner table by your own hands, the rewards exceed the toll and toil. You may be able to buy pre-packaged broilers at your local supermarket cheaper than you can grow your own, but you can't buy *better* meat.

And if you decide that you don't want to raise any more chickens, you can get out of the chicken business in a hurry. Eight weeks and they're gone, unlike the business of raising a lamb or pig, which takes five months, or a calf, which can take up to a year or more.

To sum up: 25 day-old, meat-type chicks; 200 pounds of feed; necessary equipment; a suitable building; and eight weeks of very light chores add up to 75 pounds of delicious, home-grown meat.

Check Zoning Laws

First things first! Before you even consider raising chickens in your backyard, ask local officials about zoning ordinances in your village or township. Don't become enthusiastic about raising chickens until you've cleared it with the law.

Also, if you have very close neighbors, it would be a good idea to see how they feel about your new venture. If you buy 25 straight-run (un-sexed) baby chicks, half of them may be males, and, while they won't be very loud, young roosters have been known to practice crowing, especially at daybreak. However, the promise of including your neighbors in a backyard barbecue when your birds are ready may soften their protests and sate their palates.

Read About Chickens

I suggest that you borrow or buy every book that you can find about the raising of chickens. Visit your local library and see what it has to offer. Every state in the union has at least one college or university with an Extension Service poultry specialist, and they will be more than happy to send you all the literature they have, just for the asking.

Talk Chickens

Visit every chicken grower, large or small, within a reasonable distance from your home and ask questions about raising broilers for meat. People who raise chickens will tell you where they purchased their birds, what they feed them and how much it cost; and they can give you lots of helpful tips on broiler production. In your conversations, you may run across some unfamiliar terms:

Poultry. A general term covering all domesticated birds, including chickens, ducks, geese, turkeys, guinea fowl, pigeons, peacocks and swans.

Poult. A young, immature turkey.

Hen. A mature, female chicken.

Rooster. A cock or mature male chicken.

Pullet. A female chicken under the age of one year and not fully mature.

Cockerel. An immature male chicken.

Capon. A castrated male chicken, raised to a larger size than fryers and broilers and used for roasting.

Fowl. A general term designating chickens, turkeys, ducks and geese, although the latter two are usually termed waterfowl.

Broiler-Fryer. A young chicken, male or female, under three months of age, averaging about four pounds in weight, with tender meat and a flexible breastbone.

Roaster. A larger chicken, male or female, usually from three to five months old and weighing from 5 to 8 pounds.

Stewing Hens—Boiling Fowl—Soup Chickens. Older hens that have outlived their usefulness in the egg laying category and are not suitable for broiling, frying or roasting as their meat would be too tough; thus they are stewed, boiled or fricasse'd and made edible through the process of moist heat. Don't be concerned with this because you aren't raising layers, but meat type-birds and your hens will have been eaten long before they become old and tough.

Housing

Before you buy the baby chicks, you will have to have a place to keep them. The house does not have to be fancy, but when you lift those little golden balls of fluff out of their packing crate, their new home will have to pass the severe test of Goldilocks when she tasted the porridge—not too hot, not too cold, not too damp and with no drafts—just right! You can keep your baby chicks in any shed, garage, old chicken house, barn or outbuilding that meets those requirements.

The floor can be wood or concrete. The latter would be the easiest to keep clean and sanitary. It would also be the coldest, but, as you will have several inches of litter on the floor in the form of sawdust, wood shavings or chopped straw to be used as bedding, that doesn't present a problem.

Although the chicks are tiny when they arrive and need little space, within the next eight weeks they should have at least 1 square foot of space each; now is the time to prepare for their ultimate requirements. The minimum space, then, would be 25 square feet for the 25 chickens. You will need some space for the feed and water troughs; make it large enough so that you can walk into the enclosure and turn around comfortably. Provide enough room to put a bucket of feed and water down on the floor without having to worry about crushing some little chicks. Make it easy for yourself.

If you have a building available, like a barn, but it's too large, partition it off; the southeast corner would be preferable, about 10 feet by 10 feet would be fine, but 10 by 10 is not a hard and fast rule.

Construction of a brand new chicken house, even if you did all the work yourself, could easily cost over $1,000 for materials alone, and, for a beginner who doesn't even know if he or she will like raising chickens, does not make sense at this time. For those who are sure they want to raise broilers, but have no available space, see Appendix B for plans for a small, low-cost poultry house.

If you want to provide an outside yard or range for the broilers, you could attach chicken wire to the building they are housed in. The wire should be a very heavy gauge, from 5 to 6 feet high and stapled to posts set about 6 feet apart. For extra security, a strand of barbed wire should be placed 3 or 4 inches above ground level, on the outside of the posts, and at least one strand of barbed wire strung above the top of the chicken wire. The yard could be a minimum of 5 feet wide by 10 feet long, or any dimension you care to construct in larger proportions. Cut a small door in the chicken house so that they could come and go as they pleased. This way, their feeders and watering troughs would remain inside their house.

Giving the broilers an outside yard would expose them to sunshine and fresh air, improve their vigor and the color of their feathers, but it would defeat your purpose of raising the chicks to good slaughter weight in the fastest time possible. If they were outside eating bugs or pecking at clover, it would affect your control of the broilers' feed intake and rapidity of growth. They would also be exposed to predators like raccoons, foxes and birds of prey. If time is of the essence, broiler chickens are best kept in total confinement in order for them to gain the optimum weight of 4 pounds in eight weeks.

Remember, if you use an existing building or partition off part of a barn, make sure it is draft free, with no big cracks or crannies for the wind to whistle through. If the building has windows, it is important that they can be closed securely.

In the beginning, when your chicks first arrive, you won't have to worry about ventilation as the body heat and moisture release from 25 little birds is practically nil. In fact, at this time, your main objective is to keep them warm enough so that they will survive. However, if you like raising broilers and decide to carry on with it throughout the summertime and fall, then ventilation can become most important, especially when the birds get to be five or six weeks old and are feathering out.

Chickens cannot sweat, having no sweat glands, and in extreme heat they try to cool off by opening their beaks wide. Sometimes they spread their wings and seek out the coolest spot in their quarters. A panting chicken, with its beak open and wing feathers stretched way out, leaning against a cool, damp, object, is a bird in danger of extreme heat prostration. Unless

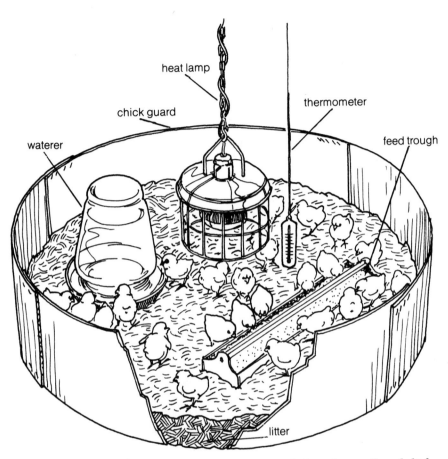

During the first few weeks of their lives, contain your chicks with a cardboard chick guard. Keep them warm with a heat lamp, suspended over the center of the area.

the bird is cooled off quickly, it's curtains. In extremely hot weather, some large poultry producers use a system of controlled ventilation or a device that generates a very fine spray at head level, having found that cool-headed birds do better. This cooling method, most often applied to mature laying hens, would not be of too much concern to a raiser of young broilers.

The housing needs of baby chicks will vary in regard to the climatic zone in which you live, particularly if you raise them in frigid weather; but this method is based on the principle of growing your very first broilers in late spring or early summer. If the enclosure you are using has a low ceiling and no windows, you may have to provide ventilation when the chicks get to be about six weeks old, with a roof vent or electric fans. As you are going to use a brooder lamp in the beginning, you must have a source of electricity in the chicken house. If there are no electrical outlets in the building, you can run a heavy-duty extension cord to the pen from the closest source.

I would suggest that you nail a fine-meshed wire on the inside of any windows so that sparrows and other wild birds cannot fly into the chicken house and spread disease.

The pen has to be secure against predators, and I include the family cat and dog in this category. You have to be able to close the door tightly at night and know that nothing can get in. There are all kinds of predators out there, from rats to weasels, who would just love to sink their teeth into your baby chicks, even if they are housed indoors. More about predators later.

When to Buy Your Chickens

The ideal time for a beginner to buy a flock of broilers is in late spring or early summer, sometime between late March and early June. At the start, you have to keep them warm and temperate weather improves your odds. It is entirely possible that one or more of your baby chicks will die within the first few weeks after they arrive. Over the years, it has been figured that the mortality rate of chickens is 2 percent for the first three weeks of their life. After that, the rate is about 1 percent per month.

Realistically speaking, this is not a high mortality rate. Considering that the baby chicks weigh little more than an ounce when hatched, their rate of respiration is extremely high; their blood circulates faster, their digestion of food is quicker and their body temperature is about 10 degrees higher than most other farm animals. They are fragile and they are delicate. However, if you follow a rigorous policy of snug housing, good feed, plenty of water and strict sanitation, you will have the satisfaction of knowing that you did the best you could. If, after all this, a couple of chicks die, it was probably inevitable. And this brings us to a very important point: buy from a hatchery of good reputation and buy the best stock that you can.

Where to Buy Your Baby Chicks

Farm journals and agricultural publications are full of the names and addresses of reputable hatcheries. These hatcheries are involved in the National Poultry Improvement Plan, which means that their breeding flocks are regularly tested for diseases by representatives of the U.S. Department of Agriculture or related state and county agencies. Some of the major feed-milling companies have annual spring sales of baby chicks, and they will stand behind their products. The Extension poultry specialist, affiliated with a college or university in your state, can help you select a hatchery. Also, the people you talked to when you were visiting local chicken farms can recommend a good source of supply.

What Breed of Chickens to Buy

There are scores of breeds of chickens to choose from in the poultry world, but we will narrow the choice down to some of the most prevalent and popular meat-type birds. We are not looking for fancy show birds, exotic bantams or high-production egg layers. Meat-type birds are not known for their egg-laying qualities and vice versa. Some breeds of chickens are excellent for dual purposes; that is, they are good egg layers and, when they've outlived their usefulness, they are still tender and meaty enough to provide succulent chicken dinners. But, for the method at hand, we will concentrate on the strictly broiler-fryer type of chicken. Among these are:

Rock-Cornish   Jersey White Giant
Cornish Super Giants   Light Brahma

Various other hybrids available to you will have names like Hubbard Broilers, Welp-Rock, Cobb Broilers, Jumbos, and Peterson Broilers. With the exception of the Light Brahma, which has dark neck and tail feathers, these are all white-feathered birds. There is a good reason for this. The meat from colored chickens tastes just as good and is just as tender as that from the all white-feathered breeds, but it all comes down to the appearance of the dressed carcass in the supermarket display counter. When white-feathered chickens are picked and dressed, the result is an attractive carcass, with its yellowish skin and lack of dark or off-color spots, caused by pin feathers. (Pin feathers are the very small, unfinished feathers that birds of broiler age often have.) Pin feathers can be hard to remove, and if they are off-color, they can detract from the overall appearance of the packaged chicken. The customer is king, or queen, if you will, and wants an attractive package to take home for dinner. And what the customer wants, the customer gets from the commercial broiler producers, some of whom turn out finished batches in excess of a million birds each week.

The consumer imposes demands on the large commercial producer as to the type of desired bird. The broiler factory, in turn, dictates the most prevalent kind of baby chick that will be offered to you. There is nothing wrong with that because you now have the best, fastest-growing, meat-type baby chick that modern science can come up with. And the broiler industry is a science, right down to the last gram of feed and rate of gain.

If you really want birds of color, you can still buy them, although they are not as common.

Buying Your Chickens

I would suggest that you buy 25 straight-run (as hatched), day-old chicks of a White Rock-Cornish broiler breed. As I mentioned, straight-run means

unsexed. That is, your chicks can be 50 percent male and 50 percent female. In the heavy-broiler breeds, the males can grow faster and utilize expensive high-protein feed better than the females. But if you request sexed cockerels, you can pay up to $3 more for 25 of them and, in the long run, you won't save much, if anything, on the feed costs with just 25 birds. If the hatchery has to take the time and trouble to determine the sex of the chicks, it costs them more money and the cost is passed on to you.

Why buy just 25 chicks? This number is small enough so that you won't get in over your head. You won't take a financial bath. Their care and feeding will be relatively easy and, when it comes time to butcher them and dress them out, you won't feel overwhelmed by the enormity of the task.

Also, most hatcheries don't sell day-old baby chicks in lots of less than 25. In transit, the chicks depend on body heat for their survival, and an amount of less than 25 doesn't make for much warmth. Then too, there isn't much profit in an order of less than 25 chicks, and the hatchery has to live, too.

Debeaking. To de-beak or not to de-beak? That is the question. And, if in your case, the decision is not quite as weighty as Hamlet's was in his famous soliloquy, it will surely affect the little chicklets you're raising in your own hamlet. What is de-beaking? It is the process of cutting off about one-half of

A properly debeaked day-old chick.

the top mandible (jaw) and blunting the lower part. If the job is done at the hatchery, it will be done by experts with a precision machine. If they use an electrical device, the wound is cauterized at the same time. It can also be done at home with a pair of wire cutters, pig-tooth nippers or heavy-duty (animal type) nail trimmers.

Why de-beak? The chicken has a decided overlap of the top part of the beak over the lower mandible. This is how they peck at things. When they eat, there is a tendency for the natural beak to hook at their feed and fling it out of the trough, thus causing waste. De-beaking cuts down on feed costs. Also, the term *pecking order* originated with chickens. The most aggressive birds pick on the more timid ones. They pick at their combs, feathers, toes and vents. Once they draw blood, that bright spot becomes a magnetic draw for other chickens to pick on the same unfortunate bird. A group of chickens can gang up on a timid bird and mutilate it. This is called cannibalism.

The hatchery will de-beak your chicks for about $.03 each, at one day of age.

The problem here is that the beak has a tendency to grow out again, and it may be necessary for the bird to be de-beaked again at nine or ten days of age. Another drawback is that a day-old chick, if de-beaked, may have difficulty in drinking water or eating food and thus may die from starvation.

In my opinion, as you are just raising 25 chicks and providing them with plenty of elbow room and they are not under any great stress or strain, as they would be if you were raising thousands at a time, I don't think de-beaking is necessary. They look terribly deformed.

For another $.03 per bird, you can have them dubbed. This means having their combs cut off. One less appendage for macho birds to peck at. Also, combs can freeze in sub-zero weather and cause problems. But, you are only going to have these chickens for about eight weeks and it will be in the springtime. I would leave their beaks and combs alone. Let them look like chickens. Let them be chickens.

Preparation for the New Chicks

Before the chicks arrive, make sure their new home (pen or shed) is swept, brushed and scrubbed. If it housed chickens before, use a good disinfectant on the walls and floor and let it stand vacant for a week or two. There are any number of disinfectants available at feed and hardware stores. Be sure to follow the directions carefully so there is no chance of chemical residue injuring the little birds.

Hang a 250-watt heat lamp in the center of the pen at about 24 to 30 inches above floor level. The lamp should have a porcelain socket, a good-sized reflector and a safety guard over the bulb. An infra-red lamp is preferred to the white-light kind. Infra-red light discourages cannibalism among chickens. Make certain that the lamp is attached securely overhead, whether it is fastened to a rafter or the ceiling. If necessary, use staples or screw hooks to secure the lamp's cord or chain. At first, you will have to go through a trial and error period to determine the correct height for the lamp. A chain would make for easy adjustments, up or down.

For the baby chicks to survive, maintain an approximate temperature of 95° F. for the first week of their lives. To do this, adjust the height of the infra-red heat lamp and constantly check the temperature of a thermometer hanging at a level about 5 or 6 inches above the floor. Any kind of inexpensive thermometer will do, but it should be attached to a sturdy backing.

The next step would be to make a chick guard. Use corrugated cardboard boxes; take them apart and cut into lengths at least 1 foot high. Staple the ends of the cardboard together and erect this barrier in a circular fashion on the floor around the heat lamp and at a distance of from 3 to 4 feet away. In other words, the heat lamp will shine down in the center of the chick guard.

This guard keeps the chicks close to the source of heat, feed and water, and can help prevent drafts at floor level. The guard has to stand sturdily so that it can't fall over and crush the chicks. It should be circular as chickens have a tendency to panic on occasion. Sometimes thunder or loud noises will set them off. If they all flee to a square corner, they can pile up and injure themselves or smother.

When the guard is in place, put a layer of coarse sawdust, wood shavings or finely-chopped straw, 3 to 4 inches high, on the floor within the guard. This is called litter and has to be dry and absorbent. It serves as bedding for the chicks and helps keep them warm. When it becomes soiled or damp, pick up the dirty areas and replace with fresh, dry litter.

As the birds grow older, enlarge the chick guard to give them more room and, of course, add more litter. After a few weeks, they outgrow the need for the guard so you can remove it, and then cover the entire floor of the pen with bedding.

Place the 1-gallon waterer and feed trough in close proximity to, but not directly under, the heat lamp.

The day before your chicks are expected to arrive, turn the heat lamp on and experiment. Leave it on for a couple of hours and then check the temperature of the thermometer, which is hanging next to and within the guard. If the temperature at floor level is over 95° F., raise the lamp a couple of notches. If too cool, lower it. Common sense.

Obviously, you must have feed on hand before they arrive.

When the Chicks Arrive

When your local post office notifies you that your chicks have arrived, open the shipping crate in the presence of post office personnel and count the live birds. Reputable hatcheries always insure parcel post shipments for loss or damage en route. In the event of damage or shortages, file a claim immediately with your post office. Some hatcheries include a couple of extra chicks to cover in-transit mortality. All of them will reimburse you for any loss you incur.

As soon as you get the chicks home, remove them from the box carefully, one by one, and dip their beaks into the waterer (lukewarm temperature) and then into the feed trough. Make sure the lip of the trough is low enough so they can reach into it easily. The trough should also be equipped with a roller bar on top so that they cannot climb inside and soil the feed. In most cases, they will learn to eat from the trough within a few days. But, in the beginning, they may be intimidated by the trough, and it would be a good idea for you to pour a bit of feed into small box tops or old egg flats and distribute these in various places within the pen, until the chicks become accustomed to eating from the feeder. Once they are eating from the trough, discard the impromptu feeders.

Feeding the Chickens

Begin by feeding a complete chick-starter ration. All of the major milling companies provide a good product. The feed will have a protein content of from 20 to 24 percent. A chick-starter ration contains corn, soy-bean meal, inorganic elements, amino acids, trace minerals, vitamins, antibiotics, and a coccidiostat, which is a medication added to prevent the disease, coccidiosis. Of all the diseases common to chickens, coccidiosis is probably the one that could most affect your birds. Chickens raised on litter or, as they say, on the floor, are more prone to coccidiosis than those raised in cages. Coccidiosis is transmitted from one bird to another by their droppings, and thus chickens raised on the floor can be exposed to the disease. The symptoms of coccidiosis are listless birds, lack of appetite and diarrhea. The mortality rate can be very high.

You will feed the chick-starter to your young broilers for the first five or six weeks of their life and then switch to a grower or finishing ration. The finishing ration is from 18 to 20 percent protein and costs a little less than the starter feed.

Most growing or finishing rations do not contain medication or coccidiostats. This is important because the birds should be withdrawn from this type of drug for at least five days before human consumption of their meat. Ask your supplier about this and read the label on the feed bag before buying a finishing ration.

Always store the chicken feed in a tight metal or plastic container so that mice, rats and raccoons cannot get at it. Garbage-type cans will do fine.

Daily Care and Observation

Set up a regular routine for feeding and watering your baby chicks. Chickens are creatures of habit, and your daily chore schedule should be consistent. When, after two or three days, the chicks are eating out of the feed trough, remove the box tops that you used for temporary feeders. Keep feed in front of them at all times, but do not fill the trough more than one-half full as they will toss out and waste a lot of feed. If they don't finish what is in the trough and it has a mouldy appearance, discard it and put in fresh feed. Empty the waterer every day and replace with clean, lukewarm water. Check the condition of the litter on the floor. If there are wet spots around the waterer, remove the damp litter with a small dust pan or shovel and replace with dry bedding.

The most important thing is to observe your chicks closely. If they are all crowded together under the direct rays of the heat lamp, it is not warm enough for them at floor level. Lower the lamp a notch or two. If they are hugging the extremities of the chick guard, as far away from the lamp as they can get, it is too hot. Raise the lamp.

Begin to reduce the temperature in the pen by about 1 degree per day, or at least 5 degrees a week, without causing discomfort to your chicks. As you observe them, they should be moving freely within their pen, eating, drinking, flapping their little wings and doing whatever it is that contented birds do.

Second Week. Remove damp and dirty litter. Add fresh bedding and stir. Everything being relative, decrease temperature to 90° F. by raising heat lamp. If the chicks appear comfortable, either enlarge chick-guard enclosure or remove it from pen. Clean waterer. Keep feed trough no more than half full.

Third Week. Lower temperature to 85° F. Observe strict sanitation procedure in pen. Cull sickly chicks. Watch for disease.

Fourth Week. Reduce temperature to 80° F. Keep litter clean and add more when necessary.

Fifth Week. Lower temperature to 75° F. Add a second feed trough. One 12-inch-long trough can accommodate 25 day-old chicks. But, by the time the birds are about five to six weeks old, they need another trough at least 12 to 14 inches long. Begin to switch feed very gradually to finishing ration. Chicks should be well along in feathering-out process.

Sixth Week. Depending on your climatic zone, gradually reduce the temperature until the birds can survive with natural heat. Don't make any abrupt changes! If conditions warrant, turn heat lamp off on warm, sunny days and put on if nights are cool.

Bright lights promote more activity, including cannibalism, among chickens. If, at this time, your chicks are comfortable without the lamp, a 40-watt bulb should provide enough light for you to do your chores and for the birds to find their water and feed, if their house has no windows. If the pen has a lot of windows, they will regulate themselves by sun time. Over-activity is not a virtue. At this stage of their lives, you want the birds to eat, drink and put on weight.

Sanitation

If you find any dead chicks in the pen, remove them immediately and bury or burn them. Keep the feed and water troughs clean. Scoop up damp spots and add fresh, dry litter. Stir the litter frequently to keep it from becoming matted. If possible, don't let strangers into the chicken house, especially those who raise chickens themselves.

Disease

Chickens are prey to many diseases, and young birds are most commonly affected by coccidiosis. The medication in their feed should help control that disease. Other prevalent diseases are fowl pox, chronic respiratory disease, cholera, blue comb, pneumonia, pullorum, typhoid and bronchitis. As a first-time raiser of chickens, you wouldn't be expected to recognize any of these. And you don't call the vet for one sick bird. Observe your chickens every time you feed them. A listless, droopy bird with no appetite is obviously in trouble. A chick with diarrhea or bloody droppings or a runny nose or eyes should be removed from the flock at once. If more than one bird shows these symptoms, consult your local Extension poultry specialist and he can direct you to a state laboratory for an accurate diagnosis of the problem and subsequent treatment. However, as your flock is small and not under a great deal of stress, if you follow a strict program of good sanitation, these problems should not affect you.

Predators

When your broilers are housed inside for their entire life span of about eight weeks, their main predators would be dogs, cats, rats and weasels. If the chicken coop is snug and tight, and the windows and doors close securely, no predators should be able to gain access. Rats and weasels can squeeze through chicken wire of 1 inch diameter. If there are holes in the pen that you suspect predators might pass through, cover them with mesh wire of not more than ½ inch diameter.

The best rat control is a system of regular bait stations, using the anticoagulant-type baits, with the poison placed along runs and walls. Traps are a good supplement if they can be placed where family pets won't get caught in them. I like cats, but have never been lucky enough to own one that was much good with rats. Mice yes, rats no. A new mother cat with nursing kittens can be tough on rats and even weasels, but it's not practical to raise litters of kittens just to control the rat population. One rat can eat 25 pounds of expensive chicken feed in a year and destroy at least three times that much. Using poison bait, supplemented with traps and a cat or two is the best way to avoid rat problems.

After Eight Weeks—Then What?

Your chicks should be approaching optimum broiler weight. You have fed them a complete ration of high-protein feed, supplied plenty of clean, fresh water and followed a steady routine of good sanitation. Pick out several birds at random and weigh them. If they weigh only 3 or 3½ pounds, keep feeding them a week longer. If they average 4 pounds, they are ready for slaughter.

Some commercial broiler factories are able to produce a 4-pound broiler in seven weeks. I think you will have done well if your birds weigh 4 pounds in eight weeks.

In 1930, it took 5 pounds of feed to produce 1 pound of broiler. By the late 1960s, this had been improved to 3 pounds of feed to make 1 pound of chicken. In this eighth decade of the twentieth century, the ratio of feed to gain is two to one. That is: 2 pounds of feed for 1 pound of live chicken. You will have to feed 200 pounds of ration for your 25 birds to gain an average of 4 pounds each. In comparison to most other farm animals, the broiler industry has made great strides in improving the feed to gain ratio.

The Day of Reckoning

Now that your broilers are of good size, it's time for you to convert them from feathered friends to meat for the table. That is the whole idea. It's possible to make pets of chickens; once I had a hen who would fly up onto my shoulder every day at feeding time. But her life expectancy was short and our friendship was brief. The cardinal rule in raising poultry or animals for meat is don't make pets of them.

The following sad tale appeared in a recent classified advertisement in my local newspaper:

"POULTRY FOR SALE—ROOSTERS AND HENS—$6.00 PER PAIR—NOT FOR EATING."

This is a common story. The owners didn't have the heart to slaughter them and thus hoped to sell them to another soft-hearted person, with no guarantee that they wouldn't end up on the table, anyway. Don't let this happen to you.

If you have 25 birds and they all look pretty much alike and they are only with you for two months, it's unlikely that any enduring relationships will develop. You have taken care of these chickens for eight weeks and now it's time for them to provide food for you.

Preparation for Slaughter

You will need a sharp axe or knife, a length of twine and a bucket or suitable container for catching the blood. Select two chickens and isolate them in a crate or box the night before they will be killed. (I suggest two so you can do a second bird immediately following the first. The first may be a bit of a learning experience.) They can have water but don't give them any feed so that it will be less messy when you dress them out. If the isolation crate has slotted floors or a wire bottom, so much the better. Then the birds cannot pick at their own manure or loose feathers.

Slaughtering the Broilers

The old-fashioned, time-honored method of killing chickens is to grasp one by its legs, hold it over a chopping block and, with a quick movement, cut off its head with an axe. Provided the chicken doesn't move at the last moment and your aim is good, it usually works. However, once I witnessed a headless bird escape from his executioner, run across the yard and into the house through an open door, jump up onto the kitchen table and expire convulsively all over the place in a most unpleasant manner. There is a better way.

You can buy a killing cone for about $4. This is a funnel-shaped restraining device that holds the bird upside down with its head out and neck exposed to the knife. Place the chicken in the cone.

A chicken suspended in a killing cone.

With one hand, grasp the head of the chicken and pull gently, but firmly, to keep a small amount of tension on the neck. With a very sharp knife in the other hand, sever its jugular vein by cutting into its neck, on the underside, slightly to the left or right and just behind the head and jaws. When the blood gushes out, you can let go of the head, or you can keep the tension on if your inner resources permit.

I suggest that you ask an experienced person to show you how to do this the first time. In any case, the blood should be caught in a container located just below the hanging bird. If you don't wish to buy a cone, tie the feet with a short piece of twine and hang the bird upside down on a nail driven into a rafter or board overhead. Restrain its head and neck as described before and proceed. When the chicken stops struggling and the blood ceases draining, cut off its head. Now it's time to pluck the feathers.

Picking the Feathers

Fill a washtub or other container with water heated to a temperature of 130° F. Grasp the now headless chicken by its feet, slosh it up and down and hold it completely under water for about 30 seconds. This helps to soften the skin and loosen the feathers. You can begin pulling the feathers off from any spot, but pluck with the grain. Pull the feathers off in the direction in which they are growing. If you go against the grain, it can result in torn and broken skin and an unsightly carcass. If the feathers don't come out easily, immerse the bird in the water for a few more seconds and try again. Keep in mind, if the water is too hot or the bird is immersed for too long, the skin will become discolored and blotchy. If pin feathers are a problem, squeeze them out, using pressure from your thumb and a dull knife.

After you have plucked all of the feathers off of the bird, there will be a number of hairs remaining on the carcass. These can be

To loosen the bird's feathers, slosh it up and down in 130° water, then hold it under for 30 seconds.

Pull the feathers off in the direction in which they are growing. If you go against the grain, you might tear the skin.

removed by the process of singeing. You can use a candle, small alcohol burner, propane torch or the open flame of a gas stove to singe these hairs. Try not to burn the skin.

Then, cut the feet off at the first joint, and prepare to dress the chicken.

Dressing the Chicken

Now that the bird has been plucked and singed, the next step is to remove the entrails. Lay the bird on its back on a counter or block, and, with a very sharp knife, make a cut about 3 inches in length in the soft area midway between the breastbone and the vent (rear end). As you make the incision, a slight pull on the skin will give your knife a firmer cutting surface. Make the cut very shallow, being careful not to pierce the intestines which lie just below the skin you are cutting.

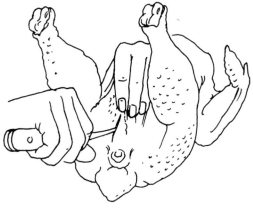

With a sharp knife, make a shallow 3-inch cut midway between the breastbone and the vent. A slight pull on the skin will give the knife a firmer cutting surface.

After making the incision, pull any remaining layers of membrane apart with your fingers. Put your hand into the incision, reach up into the body, grasp the mass of viscera, consisting of the heart, liver, gizzard and intestines, pull gently but firmly and remove from the body. At this point, the large intestine will still be attached to the body, in that it leads to the rear end and vent (anus). Make a circular cut around the vent, without penetrating the intestine, and remove entirely. Discard the intestines and place

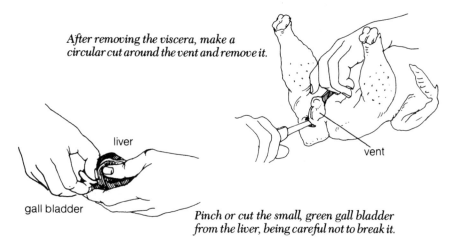

After removing the viscera, make a circular cut around the vent and remove it.

liver

vent

gall bladder

Pinch or cut the small, green gall bladder from the liver, being careful not to break it.

the heart, liver and gizzard into a pan of cold water. The gall bladder, a small, green gland, is attached to the liver. This must be removed as there is nothing in this world as bitter as gall. Cut this sac from the liver, being careful not to break it, as the contents will discolor any meat that it contacts and also cause a very unpalatable taste.

The lungs are soft and pink. If they did not come out when you removed the bird's innards, reach up into the body cavity and feel for spongy tissues next to the ribs and pull them out. The sexual glands are located next to the backbone. They are small, bean-shaped and white. They can be pinched off with little difficulty.

Cut the skin along the neck of the bird and peel it away. Locate the windpipe and throat (gullet) and separate them from the skin. Follow the throat down to the crop, a round, pouch-like receptacle that holds food the bird has eaten and serves as a primary place for the wearing down of hard grain into softer, more digestible matter. Cut below the crop to free it and then remove windpipe, throat and crop from the carcass. Cut the neck off close to the body.

There is an oil gland located just above the tail on the back side of the bird. Make a deep triangular cut around this gland and remove it. At this point, wash the carcass thoroughly, inside and out, with cold water.

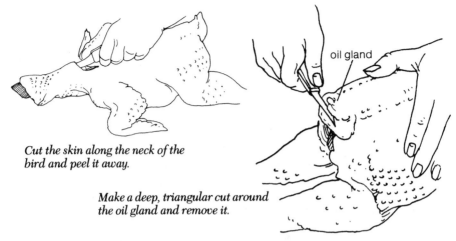

oil gland

Cut the skin along the neck of the
bird and peel it away.

Make a deep, triangular cut around
the oil gland and remove it.

Preparation of the Giblets

The giblets are the edible viscera of fowl, including the heart, liver and
gizzard. The heart and liver need only a good rinsing in cool water. To prepare the gizzard, which is a kind of second stomach for the bird, make a
shallow cut around the outer edge, through the yellow section and peel it
apart with your fingers. Discard the inside part which contains partially
ground-up food. Wash the outer halves and place with the rest of the giblets.

Make a shallow cut around the outer edge of the gizzard,
through the yellow section, and peel it apart. Discard
yellow lining and gizzard contents.

Cutting, Chilling, Packaging and Freezing

Broilers and fryers are exactly the same type of chicken, are raised in the
same way, and are the same age and weight. The difference, then, is the
manner in which they will be prepared for eating. Customarily, the broiler
is left whole, or cut in half. Fryers are usually quartered, that is, cut so that
there are two pieces combining breast and wing, and two pieces containing
thigh and drumstick. When you cut a chicken into pieces, always cut at the
joints. How do you do this? Wiggle the member you want to cut until you
can see the point at which it is joined to another piece. Then cut the con-

nective tissue between the joints. Don't massacre the bird by trying to cut through bone or muscle (meat).

It is very important that the carcass be cooled or chilled immediately after the dressing process. This prevents the development of bacteria. Obviously, a chicken cut up into several parts will cool off much faster than if the carcass (shell) is left whole.

To split the bird, make cuts along both sides of the backbone and pull the shell apart. After you have cut it in half, you can leave it in two pieces or quarter it.

Whether you leave the bird whole, split it in half, or cut into smaller pieces, begin the cooling process immediately. The main idea is to reduce the body heat of the bird to a temperature of from 34 to 38° F. as soon as possible. Cooling a whole chicken can take from six to eight hours. Chicken parts can be chilled in from three to four hours. Even if you are going to eat the bird the next day, the cooling and aging process makes the meat more tender. Chickens to be frozen should always be chilled first. You can cool the carcass in your refrigerator, or you can soak it in a bucket of ice water. If you are going to freeze the bird and use water for cooling, be sure to let it drain at least 20 minutes before packaging. The less moisture, the better it will keep. It is also important to remove as much air as possible from the wrapper when you prepare the chicken for freezing. The ultimate goal in packaging is to have the plastic bag adhere tightly to the surface of the chicken. There are several freezer bags on the market, but you can use the standard plastic wrappers available at most stores.

When the bird has been thoroughly drained of excess water and chilled to a temperature of less than 40° F., select a bag that will best suit your purpose, whether it's a whole chicken, small pieces or the giblets that are to be frozen. If you have no other devices with which to create a vacuum, insert a common drinking straw through the neck of the wrapper, draw on it to remove as much air as possible, twist the plastic several times and tie securely.

If the result of your first effort at dressing and packaging a broiler is a bit raggedy and less than professional, don't be discouraged. That's why I suggested doing two birds to begin with. Experiment with the first one. The second bird will be easier. Allowing other members of the family to participate in the operation will help considerably. You might even set up a little production line with one adult doing the killing and drawing of the bird, another cutting, and the children helping to pick the feathers and sucking on the straw to remove air from the package.

Dressing Percentage of the Chicken

Chickens dress out to an average of 75 percent. The broiler, which weighed 4 pounds when alive, will now provide about 3 pounds of edible meat, neck and giblets included. See Table 1-1.

Table 1-1.   Chickens

Approximate Weights of Body Parts		Totals
Breast	11 oz.	
Thigh	7 oz.	
Drumstick	6½ oz.	
Wings	5½ oz.	
Back	8 oz.	
Weight of shell	38 oz.	2 lb. 6 oz.

Approximate Weights of Neck and Giblets		
Liver	1¼ oz.	
Gizzard	1 oz.	
Heart	¼ oz.	
Neck	2½ oz.	
Weight of giblets	5 oz.	5 oz.

Approximate Weights of Trim		
Fat and skin		5 oz.
Head, feet, feathers, blood, entrails		16 oz.
	Total	4 lbs.

Added Benefit—Fertilizer

Over the two-month period that you raise them, your 25 broilers can produce up to 100 pounds of a very concentrated manure (litter included) that is high in nitrogen content. Although applying fresh chicken manure to flower gardens or lawns is not recommended, this manure is an excellent additive for your compost pile.

References and Sources

Animal Science, M. E. Ensminger, The Interstate Publishing Co., Danville, Illinois, 1969. A wealth of information; covers everything from beef cattle to pleasure horses; several chapters on raising broiler chickens.

Raising Poultry the Modern Way, Leonard S. Mercia, Garden Way Publishing Co., Charlotte, Vermont, 1975. A complete guide to the raising of all types of poultry; includes an excellent section on meat birds.

Broiler Industry, Watt Publishing Co., Mt. Morris, Illinois. This trade publication is aimed at those who deal with huge numbers of birds, whether as prime producers, wholesalers or retailers. However, this monthly magazine often offers articles that can be useful to anyone who raises broilers, even on a small scale.

CHAPTER 2

TURKEYS

These days, when a comedian or one of those interchangeable characters on a TV sit-com mentions the word turkey, it brings an automatic laugh from the audience. It wasn't always so. Although the domestic turkey is not known for being too clever, the wild turkey is a crafty, sly and elusive bird. Ask any hunter.

Originally native to Central America and Mexico, the turkey was fully domesticated by our southwestern Indians before Christopher Columbus landed in the West Indies. The Apache, Comanche and various other tribes used the turkey as a food and also for sacrificial occasions. In American Indian mythology, the turkey was considered to be a charitable character who provided corn for the people, outwitted the artful owl and challenged the eagle in combat. That's a pretty tall order and nothing to laugh at.

In the process of domestication, the turkey lost most of its native cunning and natural aptitude for survival. Its loss is our gain. A young, domestic turkey grows tall in short order and its growth rate is phenomenal. Consider the fact that, when hatched from the egg, the turkey poult (young bird) weighs about 2 ounces. At two weeks of age, the poult can weigh 6 or more ounces. Thus, the poult is capable of at least tripling its weight, or better, in just 14 days. That's phenomenal.

The turkey has symbolized our national holiday from the very first Thanksgiving to the present. In fact, among many people—those with a penchant for abbreviation—Thanksgiving has become Turkey Day, easier to say and more to the point.

If you would like to emulate the early pioneers and raise turkeys for your own Thanksgiving table and other holidays, here is a system for producing oven-ready birds. If you follow this method, you can produce an 18-pound tom (male) turkey, in about 20 weeks. The bird will dress out to about 14 pounds, neck and giblets included, and provide a royal feast.

This system is based on your purchasing at least six, medium-roasting-type, day-old poults, feeding them a commercial ration for about five months, and then butchering and dressing them out yourself. Obviously, you must have access to a large freezer.

The six turkey poults will cost about $18. If you don't live near a hatch-

23

ery, postal delivery charges can add up to another $5. In order for the turkeys to gain a live weight of 18 pounds in five months, the birds must consume up to 50 pounds of feed, each. Hen (female) turkeys don't eat as much as toms, but they don't grow as fast or as large, either. The six turkeys could eat about 300 pounds of feed that will cost about $35. Feed prices can vary considerably depending on the area of the country in which you live. I am quoting current prices in my locality in order to give you a general idea of what it can cost.

Equipment such as new feed and water troughs and a brooder lamp will cost another $17. If possible, attend farm auctions where this equipment may be purchased for much less. Assuming that you already have some sort of shed, garage or outbuilding to house the turkey poults, your start-up costs will be about $75. Remember, the equipment can be used over and over again, if you decide to continue raising turkeys.

You will spend about 15 minutes twice a day, every day, feeding, watering and tending your birds, or about three and a half hours a week. With supervision, a capable child of 12 or more can share the chores with you.

To sum up: six day-old turkey poults; about 300 pounds of feed; equipment such as a brooder lamp, waterer and feeder; a suitable pen or house; and about 20 weeks of moderate chores will provide you with six roast-turkey dinners. Or, if you prefer, about 75 pounds of meat* (bone included) that can be rolled, curried, fashioned into pies and casseroles, served à la king or made into a mountain of club sandwiches.

Consider these factors before you begin raising turkeys: the turkey poult costs considerably more than a broiler chick; the poult will eat about six times as much expensive feed as the broiler; the bird will be gobbling and trotting around your place for about five months; young domestic turkeys are not as hardy as chicks, ducks or geese.

Check Zoning Laws

Before you order turkey poults from a hatchery, ask your town clerk about local ordinances concerning the keeping of poultry. Even if you explain that you are only trying to add a touch of early Americana and color to the countryside, your town officials may not agree that it is your patriotic duty to raise turkeys. They may even point out the fact that the bird on our national emblem is the bald eagle, Ben Franklin notwithstanding. When the Continental Congress appointed Ben Franklin in 1776 as a member of the committee assigned the task of designing an official seal for our new country, he opted for the turkey over the bald eagle. The turkey, he thought, was a more beneficent and respectable bird than the eagle, which he considered

---

* Figure based on dressing percentage of six turkeys, half of which may be hens that will not weigh as much as toms at slaughter.

of poor moral character. Ben's choice was overruled by Thomas Jefferson and John Adams, and he lost his case.

Clear it with city hall before you begin to harbor large birds with featherless heads who go gobble, gobble all the time and sometimes don't know enough to come in out of the rain.

Read About Turkeys

You are fortunate if your local library has more than one book devoted to the subject of raising poultry, or turkeys, per se. However, every state has an Extension Service poultry specialist, usually associated with a state college or university, and I would suggest that you ask them for any literature they have on raising turkeys. It has been my experience that these specialists will go out of their way to help you with any poultry project that you have in mind. See Appendix C for a list of Extension Service offices affiliated with universities.

Visit Turkey Growers

If you don't know anyone who raises turkeys in your area, ask your poultry specialist or local feed store personnel for names and addresses of turkey keepers. Visit every turkey grower within a reasonable distance from your home and ask them how they feed and manage their birds. Talk turkey, or, in the popular idiom, get to the bottom line.

Housing

You can house the turkey poults in any type structure, whether it's an old barn, shed or chicken house, provided it's dry and draft free. The floor can be of wood or concrete, the latter being the easiest to keep clean and sanitary. Earthen floors are not recommended for turkeys, as the birds are subject to many diseases that can be caused by damp flooring underfoot. Although you are going to use dry litter as bedding, an earthen floor would tend to be damper than either concrete or wood.

If you use an old chicken house, make sure to clean it thoroughly. This means scraping the droppings and removing any old litter from the floor. You should also wash the ceiling, walls, floor and any fixtures within the house. There are many effective disinfectants on the market, and I would suggest that you consult your Extension poultry specialist for the one best suited for your purpose.

From day old to eight weeks of age, the poults will only require about 1 square foot of space each. However, by the time they are 16 weeks old, at least 3 square feet of space should be allowed for each bird. Now is the time to prepare for their ultimate requirements. The six birds will require a minimum of 18 square feet. This space, which is little more than 4 feet square, would be quite restrictive for the average adult to function freely in while doing the necessary chores. There is no reason that their pen cannot be 6

Day-old turkey poults.

feet by 6 feet, or 8 feet by 8 feet or 10 by 10 feet. In fact, they could have the run of a whole barn, shed or chicken house, because when the poults arrive, you are going to erect a draft guard or barrier of small diameter to confine them within the area warmed by the heat lamp for the first few and vital weeks of their life.

If the building you house them in is very large, you can always partition it off. The southeast corner is usually most preferable. No matter what kind of building you use, it must be draft free. All windows and doors must close tightly. I would suggest that you nail a fine-meshed wire on the inside of all windows so that sparrows and other wild birds cannot fly into the pen and spread disease.

Ventilation is a serious problem for commercial turkey growers who raise thousands of poults in confinement at a time. However, the moisture release from your six birds will be about zilch, and that's one less obstacle to worry about.

For those who want to get started raising turkeys, but have no existing, available space, see Appendix B for plans for building a small poultry house.

The housing needs of the poults will vary in regard to the climatic zone in which you live, but this method is based on the principle of starting your venture in late spring or early summer and thus taking advantage of temperate weather.

The experts maintain that if you raise turkeys, no other species of poultry or even four-legged animal smaller than a horse or cow, should be quartered on the same farm. This premise is based on the susceptibility of turkey poults to many diseases and is most applicable to those who raise large numbers of birds. If you have just six turkeys, the threat of disease decreases in relative proportion.

There is nothing like experience, however, and I would not ignore the advice of poultry experts. There are very good reasons for not raising turkey poults and baby chicks in the same pen. Turkeys are susceptible to coccidiosis and blackhead disease, both of which can be transmitted to the poults through the droppings of chickens. Also, after a few weeks, because of their larger size, the turkeys would bully the chickens at the feed trough; and very important, the feed requirements of poults differ from those of chicks.

Although I have raised turkeys, chickens, ducks, geese, rabbits, pigs, sheep, goats, calves and horses, all at the same time and on the same farm, with no dire consequences, I would not keep turkey poults and chicks in the same pen. If any poultry were housed in the building you plan to use for the turkeys, clean it thoroughly and let it stand for at least 30 days, before allowing the turkeys inside.

The turkey pen must be secure against all predators, wild and domestic, including the family dog or cat.

When to Buy Your Turkey Poults

Late spring or early summer is the ideal time for the novice to purchase day-old turkey poults. Most hatcheries only offer poults for sale from March through July. Temperate weather is a plus factor in raising turkeys. Of all the types of poultry discussed in this book, turkeys are the least hardy. For those who raise turkeys in large quantities, the general rule of thumb is to expect a mortality rate of 3 percent during the first three weeks of their lives. However, with just six turkey poults, if you follow a policy of cleanliness and good sanitation, it is possible that you can raise all of them to maturity.

Where to Buy Your Turkey Poults

Farm journals and rural newspapers are good sources of information; they often publish advertisements from poultry hatcheries. The Extension poultry specialist in your area can help you find a supplier. It is important to buy the best stock that you can. A reputable hatchery will be associated with the National Turkey Improvement Plan. This means that their breeding flocks are being continually upgraded to achieve fast, efficient growth. Also, the objectives of the plan are to reduce losses from hatchery-disseminated or inheritable diseases. Additional information about the National Turkey Improvement Plan may be obtained by writing to the Poultry Research Branch, Animal Science Research Division, Agriculture Research Service, Beltsville, Maryland 20705.

I would suggest that you buy your birds as close to home as possible. Turkey poults are not particularly hardy, and the less stress and strain they are exposed to, the better.

What Kind of Turkeys to Buy

There are several standard varieties of turkeys, but most commercial hatcheries will only offer you a choice of two types: the Broad-Breasted Bronze and the Broad-Breasted White. Both of these are strains developed from the original Bronze turkey.

Although its basic plumage is black, the Bronze turkey has feathers of varying greenish-bronze hues, suffused with coppery-bronze and white. Its warty, naked head is red, and the fleshy appendages called the caruncles (wattles and comb) turn whitish and blue when the bird becomes excited. Red, white and blue. The colors of the flag. As American as apple pie. The turkey was a staple of the pilgrims' diet in the 1700s and 1800s, but the wild bird was much smaller and had longer legs and averaged from 12 to 14

Broad-Breasted White.

pounds. Today's domestic, broad-breasted tom can weigh 25 pounds or more in just six months. The slim turkey cock of old evolved into a modern tom with dimpled breast so broad he can't fulfill his mating obligations.

Of the two varieties of turkey poults offered by most hatcheries, the Broad-Breasted White is the most popular. We have white bread, sugar, rice, rabbits, chickens, ducks and geese, so why not white turkeys? From the commercial grower's standpoint, white is better. When turkeys are dressed out and packaged for the display counter at the supermarket, the carcass of a white bird is more attractive because its pin feathers (tiny, immature feathers), when plucked, are light in color. The pin feathers of a Bronze turkey are dark and can cause skin discoloration. As you are only raising a handful of turkeys and they are destined for your own table, rather than the critical gaze of a bright-eyed shopper, this should not be an important factor in your choice of a variety.

Most important is your choice of a turkey variety that is fast growing and uses feed efficiently. Some turkey breeding farms have even reported flock records showing tom weights of 30 pounds at 20 to 21 weeks. As a small-scale grower, your chances of obtaining this type of poult are rather slim. However, intensive research and development by the large turkey breeders have greatly improved the quality of poults, particularly the Broad-Breasted White. In 1982, some 2,000 turkey farms in the United States produced about 163 million birds and the majority of them were Broad-Breasted Whites.

Buying Your Turkeys

I would suggest that you buy at least six, day-old poults of a Broad-Breasted White variety. They are quite flexible in regard to the age at which they can be slaughtered and tend to dress out well at almost any size, providing a plump, compact carcass. It would be a good idea to ask the hatchery about the growth and size potential of the poults before you place your order.

An important point: in contrast to other poultry such as chickens, ducks and geese, where there may be only a few pounds difference in the mature weight of males and females, the tom turkey grows decidedly larger and weighs much more than the hen. This difference is apparent by the time the birds are about 10 weeks old, and the gap in size increases rapidly from then on. Thus, while a tom can weigh 18 pounds at 20 weeks, a hen that same age may weigh about 12 pounds. It would not be economically justifiable to keep on feeding her in order to bring her up to a much higher weight. A hen of the medium roaster type can dress out well at a lighter weight.

If you don't live close enough to a hatchery to pick up the poults yourself, and the hatcheries you correspond with won't sell turkey poults in quantities of less than 12 or 15 birds, try to get a neighbor or friend to go in on the purchase with you and then split the order.

You may wish to save a tom and a hen from the butcher block and keep them around for local color. I would not, however, keep a pair in the hope of breeding them and then raising your own poults from scratch. The modern broad-breasted turkey is almost too clumsy to mate in the natural fashion; the majority of turkeys are bred by artificial insemination.

Preparation for the New Poults

The pen or building that the poults are to be housed in must be as clean as possible. This means sweeping the ceiling to remove cobwebs, washing the walls and sills and scraping any dirt or old litter from the floor. Seek your Extension Service poultry specialist's advice in regard to disinfecting the premises, and follow the manufacturer's recommendations when using poisonous chemicals.

For the poults to survive, they must be kept warm and dry. You will have to supply artificial heat for the little birds for the first few weeks of their lives. To do this, hang a 250-watt heat lamp in the center of their pen, about 18 to 24 inches above floor level. The heat lamp should have a porcelain socket, reflector and a metal guard over the bulb, in case it falls by accident. Infra-red bulbs are preferable to the white-light kind.

You will have to try and maintain a constant temperature of 100° F. in the poults' pen for the first week.* To do this, you must be able to adjust the height of the heat lamp which you will hang overhead by a rope or chain. To check the floor temperature, hang a sturdy thermometer about 5 or 6 inches above the floor and at least 3 feet away from the heat lamp. Experiment with the heat lamp, raising and lowering it, until you find the correct height at which you can maintain the desired floor level temperature of 100° F.

The next step is to erect a draft protection guard or shield. This guard should encircle the area warmed by the heat lamp and be at least 3 feet away from its rays. The guard should be at least 12 inches high. You can buy one for about $3.50 or make your own from corrugated cardboard boxes. This shield not only protects the young birds from drafts, but also confines them to the source of feed and water in the beginning. It is circular because when turkey poults are startled by loud noises or electrical storms, they have a tendency to panic and play follow the leader and all run in one direction. If the shield has square corners, they could pile up and smother or injure themselves.

When the guard is in place, put down a layer of wood shavings, peat moss or straw, from 3 to 4 inches high, to serve as bedding. The bedding or litter, which must be dry and absorbent, helps keep the poults warm. Sawdust is not recommended because, if damp, it can become mouldy and cause dis-

---

* The United States Department of Agriculture recommends that Broad-Breasted White turkey poults be brooded at about 100° F. for the first week. The recommended temperature for Bronze turkeys is 95° F.

ease. Do not use newspapers or any slick-surfaced material, as the poults can develop leg problems if they slip on a smooth surface. Obviously, the hanging thermometer that you are using to check the floor temperature must be within the confines of the draft guard.

After a week, you can enlarge the diameter of the guard, adding more bedding so that the poults don't stand on a bare floor.

Place a 1-gallon waterer and feed trough close to but not directly under the heat lamp. The water fountain should not be wide enough for them to climb into, as they could become chilled and die. The feed trough should have a spinner bar on top of it so that the birds cannot climb inside and soil the food.

The day before the poults are due to arrive, turn on the heat lamp and check the temperature at floor level. If it's above 100° F., raise the lamp. If it's below the desired temperature, lower the lamp. You will have to go through a process of trial and error to maintain the correct degree of warmth.

Make sure you have feed on hand before you bring the birds home.

When the Poults Arrive

If your poults come through the mail, open their shipping container in the presence of a postal clerk and count the birds. Parcel post shipments are always insured by reputable hatcheries in the event of loss or damage occurring en route. If there are any injured or dead birds, file a claim with your postmaster and send it to the hatchery.

As soon as you get the poults home, take them out of the crate, one by one, and dip their beaks into the fountain, which should be filled with luke-warm water. Then introduce them to their food by dipping their beaks into the feed trough, which should be filled to the brim. Turkeys are notoriously slow learners, and you may have to repeat this process several times a day for the first few days before they catch on. Poults have been known to die of starvation while standing on their feed. With most poultry, it is not recommended to fill feeders more than about three-fourths full, because of the chance of waste as the birds fling feed out of the trough. However, in the case of turkeys, it's better to lose a little feed than a young bird, and a full trough might just help a poult get his act together more quickly. Some growers place shiny objects like bright-colored marbles in both the water fountains and feed troughs in order to attract the poults' attention. It would be unkind to say that turkeys don't have all their marbles and this is one way of resolving that lack.

You can also put small amounts of feed in little box tops and place them in various locations near the heat lamp to help the poults learn how to eat. It is very important that you observe the birds closely, rather than just doing the chores mechanically and by rote. If they are huddled under the heat lamp, lower it an inch or so. If they are crowding against the draft guard, raise the lamp. If they don't know how to eat or drink, show them.

Contented turkey poults will move freely within their pen, eating, drinking and trying to untie your shoelaces.

Feeding the Poults

The turkey poult's body is formed entirely from the egg and shell during the incubation process. Just before it hatches, the remaining egg yolk is ingested into its body. The yolk provides food and liquid for the bird during the first hours of its life. Accordingly, hatcheries are able to ship poults for long distances. However, the sooner you feed and water them after hatching, the better their chances for survival.

To begin with, the poults should be fed a turkey-starter ration of about 28 percent protein, in mash or granule form. The feed will be medicated and contain a coccidiostat, which helps prevent the disease, coccidiosis. This disease can be fatal for poults. Some imaginative turkey raisers place a few baby chicks among their birds for the first few days to teach the poults how to eat and drink. Chicks are fast learners. But, as chicken droppings can cause coccidiosis in turkeys, I would not recommend this practice. Blackhead is another destructive disease which can affect turkeys of all ages. If your turkey-starter ration does not contain a blackhead-static or antihistomonad drug, consult your Extension poultry specialist or a veterinarian knowledgeable about poultry. Ask them for a suitable medication that can be added to the turkeys' drinking water.

You will feed the starter ration to the poults for the first eight weeks of their lives. After the birds have learned how to eat, I would only fill their feeders about three-fourths full, but always keep feed in front of them—free choice. As a general rule, the lip of the feed trough should be the same height as the bird's back. In the case of day-old poults, the sides of the trough will be about 3 inches high. By the time they are eight weeks old, the sides of the trough should be from 6 to 8 inches high. A feeder is just a rectangular box with four sides and an open top, and you can construct a larger one using lath or scrap boards. Always use a trough. Do not spread their feed on the floor or ground.

After eight weeks, switch the poults to a growing ration of about 21 percent protein. This ration will be fed to the birds until they are about 16 weeks old. From then on, until they are 20 weeks of age and approaching optimum slaughter weight, you can feed them a finishing ration of 16 percent protein. In the beginning, they need a great amount of protein, as they grow so fast. If they don't have enough protein, they can develop leg problems or become lame. As they get older, their protein needs decrease and their energy requirements increase.

Those who raise turkeys in large numbers supplement feed rations with grains like corn, wheat or oats, thus saving on feed costs. It would not be practical to buy a sack of corn for just six birds.

From day-old to eight weeks of age, the six poults, if of mixed sexes, will eat about 50 pounds of feed. From 9 to 16 weeks, the birds will consume about 150 pounds of feed. From 17 to 20 weeks, they will finish off about 100 pounds. Theoretically, then, you will need one bag each of starter, grower and finisher. Commercially pre-packaged rations come in 100-pound bags. After eight weeks, when you are ready to switch to turkey grower, you will have a considerable quantity of starter left over, perhaps 50 pounds. You can mix this with the grower, when you make the change.

Growing and finishing rations do not usually contain medication. Poultry feed containing drugs should be withdrawn from the birds for at least five days before human consumption of their meat. If there is any question about this, read the label on the feed bag and ask your supplier.

I would suggest that you supply grit for the poults, in order for them to develop their gizzards. The function of the gizzard is to grind food and help the digestive process. You can buy insoluble grit at a feed store or save money by using very fine gravel stones. Always store feed in tight containers so that rats, mice, squirrels and raccoons cannot get into it.

The turkeys can be confined to their pen for their entire lives. Total-confinement rearing affords protection against predators, thieves and inclement weather. The main disadvantage is the possibility of respiratory diseases, particularly caused by indoor confinement.

Turkeys are good foragers, but for just six birds, it would not pay you to plant a crop of soybeans, kale, rape or alfalfa for them to range on. By the time the poults are eight weeks old, they should be well feathered out. It would be ideal if you could arrange for them to have access to a small yard for exercise, fresh air and sunshine. Weather permitting, you would let them out during the daytime and pen them in at night.

The yard should be of sandy soil or good sod, be well drained and have no stagnant pools of water within. The dimensions of the yard will depend on the space you have available. For six birds, about 50 square feet of yard-space will suffice. You would provide shade and drinking water and feed them from a trough, not spreading feed on the ground. Fence the yard with a sheep or hog-type woven wire, from 4 to 5 feet high; have a strand of barbed wire at ground level and two strands at the top, to protect the birds from large predators like foxes, coyotes and dogs. A very heavy-gauge poultry wire about 6 feet high, again with barbed wire at top and bottom, will do the job. The bottom strand of barbed wire, at ground level, should be located on the outside of the fence.

Raising the Poults: the First Eight Weeks

In the beginning, it is vitally important that the young birds be kept warm and dry. For the first week, maintain a temperature of 100° F. in the poults' house or pen. After that, decrease the temperature 5° F. weekly, all the while observing the birds' comfort very closely. Lower the temperature by raising the heat lamp an inch or so and checking the thermometer hanging at floor level. This is a process of trial and error and will take a little time. Within seven weeks, the temperature in the turkey house will be 70° F.; at this point, unless your locality is experiencing a cold spell, you can discontinue using the heat lamp.

After a couple of days, if and when the poults are eating from their trough, you can remove the little box tops and lids that you used as impromptu feeders at the start. For the first eight weeks, feed them a starter ration of about 28 percent protein. The feed should contain medication for the prevention of coccidiosis and blackhead disease.

Keep their trough three-fourths full of feed at all times. If there is feed left over and it becomes mouldy, get rid of it before they have a chance to eat it. Keep the water fountain full. Don't just add more water; rinse the waterer twice a day and fill with fresh water.

Everything being relative, you can enlarge the draft guard after a few days, to give the poults more room. Be sure to cover bare spots with dry litter. If the turkeys' house is snug and tight, you can remove the guard after about 10 days.

Turkeys produce a very wet manure. Using a small dustpan or shovel, remove wet or soiled litter daily, and replace with fresh, dry bedding.

If a poult dies, take it out of the pen immediately, and bury or burn the carcass. If more than one bird becomes droopy, listless, lacks appetite or has a foul-smelling discharge, consult your Extension poultry specialist for proper treatment and advice. Medication, however, is no substitute for good management.

Raising the Poults: from Nine to Sixteen Weeks

By this time you will have discontinued the heat lamp. You can provide a dim light in their pen at night, using a bulb of low wattage. This will encourage growth and enable them to find their feeder and water fountain. The birds will be well feathered out. The toms will weigh about 5 pounds and the hens about 4 pounds. It's time to switch to a turkey-grower ration of about 21 percent protein. Mix any leftover starter feed with the grower. Observe strict sanitation and good housekeeping rules. Remove wet and soiled bedding daily, and add fresh litter, stirring occasionally. Keep feeder three-fourths full and rinse waterer when adding fresh water. If possible, erect a fence as previously described and turn the birds out into the yard for sunshine, exercise and fresh air. If penned outside, provide shade, water and feed from a trough. You can herd them back into their house at night.

Raising the Poults: from Seventeen to Twenty Weeks

Switch the birds to a finishing ration of about 16 percent protein. Keep on feeding them free-choice and always supply plenty of fresh water. Keep their bedding clean and dry.

When the birds are about 20 weeks old, catch a couple of the largest and weigh them. To catch a turkey, herd him into a dark room and grab him by

Broad-Breasted Whites on range.

both legs. Fasten the legs together with a metal hook or twine and hang him upside down onto a scale. Don't let heavy turkeys hang for more than 2 or 3 minutes. If the toms weigh about 18 pounds and the hens about 12 pounds, you are right on target. The carcass should feel plump with no prominent bones protruding. To check degree of finish, select a place in the sparsely-feathered breast area and pinch a fold of skin between the thumbs and forefingers of both hands. The skin should feel thick and be of an off-white, creamy color. Under-finished birds have paper-thin skin, almost transparent and of reddish hue. If the birds are of good weight and the skin feels right, sharpen your axe. If they are light and bony, keep on feeding them for another week or two, until they reach the desired condition of finish. Be sure to withdraw any medicated feeds from their diet five to seven days before slaughter.

Disease

Turkeys are about the least hardy of all domestic fowl and are subject to many diseases. But, you should have no disease problems, as you are raising only six birds (under a strict program of cleanliness and sanitation), isolated from other types of poultry. If more than one poult appears sick, though, consult your Extension poultry specialist for direction.

Predators

Housing the poults within a secure pen for the first two months of their lives will prevent any loss by predators. It is up to you to keep pets like the family cat and dog under control. Weasels and rats can squeeze through a space about 1 inch in diameter, and any sizable holes in the turkey house should be covered with fine wire mesh or boarded up to keep predators out and prevent drafts. Usually, rats will not bother poults after they are a couple of weeks old. However, one rat can eat or otherwise destroy up to 100 pounds of expensive feed in a year. Don't support rats. Remove any litter or old trash piles from the turkey yard area, and don't give rats any place to breed or hide. Anti-coagulant baits, supplemented with traps, are a good method of rat control. If you are able to provide a yard for the turkeys after they are feathered out, penning them in their house at night should prevent losses from the likes of foxes, coyotes and stray dogs.

After Twenty Weeks

Turkeys are interesting and curious birds, and the longer you keep them, the curiouser they become. Don't make pets of them and try not to talk to them because they will usually answer you back. You have spent a lot of time and money on these birds, and now it's time for them to provide meat for your table. If the toms and hens are of good weight and well fleshed out, it's time to steel your heart, gird your loins and go for it.

Killing, Plucking, Evisceration and Preparation for Freezing

I would suggest that you contact a person experienced with the whole process and have them show you how to do it, step by step, from start to finish. The killing, featherpicking, eviscerating and preparation for freezing a turkey is the same as doing up a broiler chicken, except that you are now dealing with a bird about five times larger. (See Chapter 1 on chickens for guidance.) White turkeys have more feathers than their bronze brothers. Before plucking they should be subscalded in water (140° F.) for from 30 to 60 seconds.

When picking the turkey, pluck the tail, wing and leg feathers first, save the breast and neck till last. If the feathers do not come out easily, immerse the bird again in the hot water and start over. According to the *Guinness Book of World Records,* the best recorded time for plucking a turkey stands at 2 minutes and 44 seconds. This record was established by an Irishman in a British Isles contest. If it takes you a half hour to pick the bird clean, don't be discouraged. The second one will be easier.

You will eviscerate, chill and package the turkey in the same method as you would process a broiler chicken. The system is the same, although you are now dealing in larger quantities of offal and flesh. The freezer bag required will be much bigger than that used for a broiler chicken.

Dressing Percentage of Turkeys

The turkey produces a higher proportion of edible meat, in relation to live weight, than any other species of bird or animal. It averages about 78 percent when dressed out. The 18-pound tom will dress out to about 14 pounds, neck and giblets included.

Added Benefit—Fertilizer

The accumulation of eight weeks of highly concentrated manure, litter included, from the pen your six poults were housed in, will be an excellent additive for your compost pile. Adding fresh poultry manure to your lawn or garden is not recommended.

References and Sources

Animal Science, M. E. Ensminger, The Interstate Publishing Co., Danville, Illinois, 1969. Several excellent chapters on the care and feeding of turkeys.

Turkey Production, Stanley J. Marsden, Agricultural Handbook No. 393, Agricultural Research Service, United States Department of Agriculture, Washington, D.C., 1971. This handbook of 77 pages contains invaluable information for any potential turkey grower.

Raising Poultry the Modern Way, Leonard S. Mercia, Garden Way Publishing Co., Charlotte, Vermont, 1975. This up-to-date and most informative guide to the raising of all kinds of poultry contains an excellent chapter on turkey production.

Raising Your Own Turkeys, Leonard S. Mercia, Garden Way Publishing Co., Charlotte, Vermont, 1981. This 140-page book by an expert poultryman covers everything you need to know about raising turkeys. Step by step, with how-to information, the book tells you how to get started, feed and care for your turkeys, and even provides instruction on carving the finished bird.

Backyard Poultry, Rte. 1, Waterloo, Wisconsin. Monthly magazine aimed toward the small-flock owner; features articles on most types of poultry, including turkeys.

CHAPTER 3

DUCKS

One of the songs that soldiers sing as they practice marching on the drill field goes something like this: "Be kind to our web-footed friends, for a duck may be somebody's mother. . . ." Filial sentiment aside, it pays to be nice to ducks because one duckling, if properly cared for, can grow to a live weight of 7 pounds in just seven weeks and provide you with a net of 5 pounds (neck and giblets included) of delicious meat that can be roasted, barbecued or even fried.

The duck's legs and feet were made for swimming and diving. Walking on land, ducks have a tendency to waddle, their short legs being placed far to each side of their body, and when they move, their feet tend to go inward toward their center of gravity, producing the clumsy motion. If they can't perform a close-order drill with any semblance of gracefulness, they can certainly grace a dinner table and offer exciting fare.

If you would like to vary the main course on your dining table, whether for Sunday dinners, holidays or special occasions, here is a method for producing oven-ready duckling. If you follow this system, you can net about 75 pounds of meat within seven weeks, enough for at least 15 dinners for a family of four.

This method is based on purchasing 15 day-old ducklings of a fast-growing, meat-type breed, feeding them a commercial ration for about seven weeks, and then butchering and dressing them out yourself. You should own or have access to a large freezer.

The 15 ducklings will cost about $20, and, if you don't live reasonably close to a waterfowl hatchery, postal delivery charges can add up to another $5. In order for them to gain a live weight of 7 pounds in seven weeks, you will have to feed the ducklings approximately 280 pounds of high-protein starting and growing rations, which will cost about $37. Equipment such as waterers, feed troughs and a brooder lamp will cost another $28. Assuming that you don't have to build a structure to house the ducklings, but already have some sort of building such as a shed, barn or utility house suitable for raising a small flock, your start-up costs will be about $90. If you decide to continue raising ducks after the first season, the equipment can be used over and over again.

You will devote about 15 minutes twice a day, every day, to feeding, watering and observing your birds; or about three and one-half hours a week. It is a seven day a week job. You can, however, share the chores with a capable child of 12 or more. Raising waterfowl has long been a favorite project with young children in 4-H clubs. It *is* a responsibility, as the lives and well-being of 15 tiny birds will be in your or their hands. If your children's only close encounter with web-footed birds is a well worn rubber ducky in the bathtub, or the neatly packaged frozen duckling found on the supermarket shelf, it would be a good way for them to learn where our food really comes from and how it grows. The little ducklings, covered with golden down, are irresistible, but do not make pets of them. Eventually, they will provide food for your table. All things considered, you may be able to buy pre-packaged duckling at the store cheaper than you can grow it yourself, but it will never taste as good as one you raised.

If you find that you don't like raising ducks, you can be out of the duck business in short order. Seven weeks and they're gone. Not at all like the business of raising a pig or lamb, which takes about five months.

To sum up: 15 day-old ducklings; 280 pounds of feed; necessary equipment such as feeders and waterers; a suitable pen or house; and seven weeks of very light chores add up to about 75 pounds of delicious, home-grown feasts.

Check Zoning Laws

Before you even consider raising ducks in your own backyard, check with your town clerk in regard to local ordinances dealing with the harboring of poultry in your locality. Unlike his crowing counterpart, the rooster, the male duck (drake) speaks softly and in a low, throaty voice. But the female

ducks, especially White Pekins, a most popular breed of meat-type duck, can be very loud and harshly insistent in their vocalizing as they grow and mature. I would suggest that you sound out your nearest neighbors in regard to their feelings about your proposed venture. Contrary to popular belief, it is not necessary to have a fresh water pond to raise ducklings. Although the wild mother duck leads her new brood to water shortly after they are hatched, you will keep your little flock of domestic ducklings under shelter for at least the first 30 days of their lives, until their backs begin to be covered with protective feathers. The last thing you want is for them to get soaked and chilled by cold spring rains. In the wild, when a storm threatens, the mother duck calls her brood, and they scurry to seek shelter under her protective wings. Your ducklings, having no such vigilant maternal presence on hand, will have to depend on the snug roof of their house for protection from the elements.

Read About Ducks

Admittedly, the library of source books on the raising of waterfowl is slim. I would suggest that you contact the Extension Service poultry specialist in your area, every state has one associated with a college or university, and ask for any literature they have on raising ducks. Without exception, I have found them to be most helpful people. (See Appendix C.)

Visit Duck Growers

Pay a call on every duck raiser within a reasonable distance from your home and ask questions about the care of ducklings. They can tell you where to buy good birds and how they manage their flocks. In your conversations they may use some of these terms:

Poultry. A general term covering all domesticated birds, including chickens, ducks, geese, turkeys, guinea fowl, pigeons, peacocks and swans.

Waterfowl. A term designating birds that swim, i.e., ducks and geese.

Fowl. A general term referring to chickens, turkeys, ducks and geese.

Duck. In general, the family of swiming birds, ducks; in particular, the female duck as distinguished from the male of the species.

Duckling. A young or immature duck of either sex.

Drake. A mature male duck.

Hen. A mature female duck.

Housing

Before you buy the baby ducklings, you will have to have a place to keep them. Their shelter does not have to be elaborate, but in order for them to survive it has to be dry and draft free. You can house the ducklings in any shed, chicken house or barn that meets these requirements. The floor can be made of wood or concrete, although concrete would be the easiest to keep clean and sanitary. Concrete is cold, but, because you are going to have several inches of litter in the form of peat moss, wood shavings or chopped straw to serve as bedding, that will not present a problem.

Although the day-old ducklings are tiny when they arrive and need little space, by the time they are six or seven weeks old, they should have at least 2½ square feet of space each, and now is the time to prepare for their ultimate requirements. The minimum space for 15 birds would be 37½ square feet. You will also need room for the feed and water troughs and enough space within the pen so you can move about comfortably without stepping on the ducklings. Make chore time easier; give yourself a little extra room.

If there is a barn available, but it's too large, you can partition it off. If you have a choice, the southeast corner is usually most preferable, and an 8 x 10 or 10 x 10 foot section would provide plenty of room. It would not be practical to construct new housing for the ducks, even if you did all of the work yourself. Materials could easily cost over $1,000, and, for a beginner who doesn't even know if he or she will like raising ducks, it does not make sense at this time.

Whether you use an existing building or partition off part of a barn, make sure that it is draft free. Any windows in the structure must close tightly. When your ducklings first arrive, you won't be concerned about ventilation as the body heat and moisture release from 15 little birds is practically nil. In the beginning, your main objective is to keep them warm enough so they will survive. However, if they are still confined to an indoor pen by the time they are five or six weeks old, and they are feathered out, and the pen is of minimum size, 40 or 50 square feet, then ventilation could become a problem, depending on the time of year. If their pen has a low ceiling and no windows, you may have to provide ventilation with a roof vent or electric fans.

The housing needs of the ducklings will vary in regard to the climatic zone in which you live, but this system is based on the principle of raising your very first ducklings in late spring or early summer to take advantage of temperate weather. To start off, you will use a brooder lamp to keep the ducklings warm; if the building you choose to house them in has no electrical outlets, you can run a heavy-duty extension cord to their pen from the closest source of power.

It would also be a good idea to nail a fine-meshed wire on the inside of all windows so that sparrows and other wild birds cannot fly inside the house and spread disease. The pen has to be secure against all predators, whether wild or domestic, including the family dog or cat, who can wreak havoc among ducklings. Obviously, all doors must close tightly.

When to Buy Your Ducklings

Late spring or early summer is the best time for the beginner to purchase day-old ducklings. As a matter of fact, with the exception of the White Pekin, most breeds of ducklings are only offered for sale from March to July. In the beginning, the main idea is to keep them warm, and mild weather will aid in the success of your project.

Unlike baby chicks, with a mortality rate of 2 to 3 percent within the first few weeks of their lives, the duckling is exceptionally hardy, and it is entirely possible to raise every one to maturity, barring physical accident. And yet, strangely enough, the duckling shares many of the same physiological traits as the chick: a rapid rate of respiration and blood circulation; relatively fast digestive processes; and a body temperature of about 107° F., almost 10 degrees higher than other four-legged farm animals.

Even though the duckling is known for having a sturdy constitution, it is still very important to buy the very best stock that you can from a reputable hatchery.

Where to Buy Your Ducklings

Farm journals and agricultural magazines provide names and addresses of waterfowl hatcheries. In season, the classified advertisements in rural newspapers are a good source of suppliers. And in the spring, many commercial feed companies have annual sales of poultry, including ducklings. Your local Extension poultry specialist can help you choose a hatchery. Last but not least, the poultrymen in your area whom you visited when you first became interested in raising ducklings can recommend a good source of supply. There are advantages in buying your ducks locally. You will save on delivery charges and if there are problems, you can always go back to the people you bought them from.

What Breed of Ducks to Buy

There are many breeds to choose from in the duck world. Some, like the White Crested, have an ornamental tuft of feathers atop their heads. Others, such as the Khaki Campbell, can lay an egg a day, every day, for a year, and that's the kind of production that even the highest producing strain of chickens cannot come close to. However, for our purpose, we will narrow the choice of birds down to the meat types. The most prevalent and popular of these are:

White Pekin   Muscovy
Rouen

For the past several decades, the production of ducks for meat in the United States has averaged a little over 10 million birds, annually. Better than 60 percent of these are raised on Long Island, New York, hence the name, "Long Island Duckling," that appears on the menu of your favorite restaurant.

White Pekin. By far, the great majority of ducks raised by commercial growers are the White Pekin. The poultry industry and modern science have combined to develop the White Pekin specifically for the meat bird market.

They are very fast growing, averaging 7 pounds live weight in seven weeks, under good management. Their feathers are all white and their skin is yellow, which makes for an attractive carcass when they are packaged for sale at the market. When ducks are plucked, they have small, immature feathers called pin feathers, which are difficult to remove. The pin feathers of a white-feathered bird don't discolor the carcass as the pin feathers of a duck with dark or colored plumage tend to do.

Rouen. The Rouen is a very attractive bird, having the same vivid coloration as its wild ancestor, the Mallard. The mature Rouen drake has a brilliant green head, a white collar around the neck, a brownish breast and a patch of blue on each wing. In contrast, the coloration of the female Rouen, with the exception of the blue patches on her wings, is a subdued shading of browns, with thin stripes of black on her breast.

While the Rouen will grow to twice the size of their wild ancestor, the Mallard, it has been my experience that they do not reach optimum slaughter weight as quickly as the White Pekin.

Muscovy. The Muscovy is a bird of another feather. As a child, I used to call them "turkey-ducks," probably because the patches of coarse, red skin on their faces reminded me of gobblers. I always thought they were from Russia, linking the name to muscovites. In fact, they are native to South America.

The most prevalent varieties of Muscovy available are the White and the Colored, which has a combination of white and greenish-black feathers. Muscovy ducks are slower growing than either White Pekins or Rouens and would not be suitable for the purpose at hand. They are excellent foragers, however, and would be an ideal breed to turn out on range to feed on insects and seeds and grow at their own chosen speed.

Roast Muscovy makes for a delicious dinner, their meat being a bit leaner than that of the Pekin or Rouen. I can honestly attest to this fact because one fall I ate every last Muscovy on the farm. Being fairly good fliers, they had flown into the orchard, and I preferred to harvest the apples my way. In the end, the apples were delicious, and so were the Muscovies, and I had it both ways.

Buying Your Ducklings

I would recommend that you buy 15 day-old, as-hatched White Pekin ducklings. As-hatched means un-sexed. Your birds will be a mixture of male and female. Most hatcheries do not offer a choice of sexed ducklings, or if they do, they charge another $.50 per bird for their trouble. While males will utilize the high-protein feed more efficiently than females, with such a small number of birds, you will not save on feed costs by specifying male ducklings in your purchase order.

With just 15 ducklings, you won't get in too deep, financially. Taking care of them will be relatively easy, and when they are ready for butchering and dressing out, which, according to this system, you should do yourself, you won't be overwhelmed by the job at hand.

Preparation for the New Ducklings

Before the ducklings arrive, make sure their new home, whether it's a shed, pen or partitioned-off barn, is brushed, swept and scrubbed clean. If it housed poultry before, use a good disinfectant on the walls and floor and let it stand vacant for a week or 10 days. When using the disinfectant, follow the directions carefully so there is no chance of the little birds being injured by poisonous chemical residue.

In order to brood the ducklings and keep them warm, hang a 250-watt heat lamp in the center of the duck pen at about 18 to 24 inches above floor level. The heat lamp should have a porcelain socket, a sizable reflector and a metal safety band over the bulb. The infra-red type of bulb is preferable to the white-light kind. Infra-red light discourages cannibalism among poultry. Cannibalism occurs when an aggressive bird picks at another one, pulling out feathers and drawing blood. Other birds will then gang up on the unfortunate victim and can, in time, mutilate it or even cause its death. Many factors are thought to cause cannibalism, including over-heating, over-crowding, incorrect diet, lack of water and too much light. As you are only raising 15 ducklings, and you will provide plenty of floor space, a constant supply of water and a complete duck feed, cannibalism should not be a problem with your small flock.

For the ducklings to survive, you will have to maintain an approximate temperature of 90° F. in their pen for the first week of their lives. To accomplish this, you will have to adjust the height of the heat lamp. You can hang the lamp overhead by a rope and use pulleys, or you can hook it to a chain. As you experiment with the lamp, constantly check the temperature with the aid of a thermometer hanging about 5 or 6 inches above the floor. You can use any kind of inexpensive thermometer, but it should have a sturdy backing. To begin with, use a trial and error process of raising and lowering the lamp, to determine the correct height and maintain a floor temperature of 90° F.

The next step is to make a guard for the ducklings. Use corrugated cardboard boxes; take them apart and cut into lengths about 1 foot high. Staple the ends of the cardboard together and erect in a circular fashion, on the floor about the infra-red bulb and at a distance of from 3 to 4 feet away. In other words, the heat lamp will shine down in the center of the guard.

This guard will keep the ducklings close to the source of heat, feed and water, and help prevent any drafts at floor level. The guard has to stand up by itself and be sturdy enough so that it won't fall over and hurt the ducklings. It should be circular because ducklings can panic when they hear loud noises or thunderclaps, and they have a tendency to run in one directon. If the guard has square corners, they could all pile up and become injured or smother.

After the guard is in place, put down a layer of coarse sawdust, dry peat moss, wood shavings or fine, chopped straw from 3 to 4 inches high on the floor within the guard. This litter must be dry and absorbent; it serves as bedding for the ducklings and keeps them warm. When the litter gets wet or soiled, pick up the dirty areas and replace with fresh, dry bedding.

When the birds grow older, you will have to enlarge the guard to give them more space and, obviously, add more bedding. After a couple of weeks, when the ducklings have outgrown the need for a guard, you can remove it and then cover the entire pen floor with clean, dry litter.

Place the 1-gallon waterer and feed trough close to, if not directly under, the heat lamp.

The day before your ducklings are due to arrive, turn the heat lamp on for a couple of hours and then check the temperature at floor level, via the hanging thermometer. If the temperature is above 90° F., raise the lamp an inch or so. If it's 85° F., lower it. Experiment. The idea is to try and maintain 90° F.

You must, of course, have feed on hand before you bring the ducklings home.

When the Ducklings Arrive

When the post office notifies you that your ducklings have arrived, open their shipping crate in the presence of a clerk and count the live birds. Parcel post shipments are always insured by reputable hatcheries in case of loss or damage en route. In the event of damage or dead birds, file a claim immediately with your postmaster and forward it to the hatchery.

As soon as you get the ducklings home, take them out of the box, one by one, and dip their beaks into the water, which should be lukewarm. Then gently place each little bill into the feed trough, which should be equipped with a spinner bar on top, so that the ducklings cannot climb inside it and soil the feed. Within a few days the ducklings will learn to eat from the trough, but in the beginning it would be a good idea to pour a bit of feed into small box tops and distribute them in various places near the heat lamp, within the pen, until they become accustomed to eating from the feeder. Once they are eating from the trough, discard the makeshift feeders.

The waterer should be deep enough so the ducklings can submerge their entire bill. Their nostrils are located way up the bill near the head, a very clever arrangement enabling them to drink or sieve food from water and breathe at the same time. However, the water fountain should not be wide enough for them to climb into, as they can become chilled or drown. Ducks require more water than chickens; it keeps their bills, nostrils and throats clear, and helps them avoid choking on dry feed.

Feeding the Ducklings

All poultry, including ducklings, can exist without food or water for a couple of days after hatching because they absorb the yolk of the egg before they break out of the shell. This nourishes the baby bird and enables him to survive the trip from the hatchery to your place.

In the beginning, you should feed them a complete duck-starter ration with a protein content of about 22 percent. All of the major milling companies provide a good product. The ration will contain corn (or wheat), soybean meal, alfalfa meal, amino acids, inorganic elements, trace minerals and vitamins.

Do not feed them a medicated chick-starter ration, as this will contain a coccidiostat, a medication for the prevention of the disease, coccidiosis, in baby chicks. A coccidiostat in heavy doses can be fatal to ducklings, who are not usually susceptible to coccidiosis. Ducklings, however, need much more niacin in their diet than baby chicks do, and a complete duck-starter ration will contain the correct amount of this vitamin.

If you cannot purchase a duck-starter ration at your local feed store, you can use an un-medicated chick, turkey or game-bird starter feed for your ducklings. Read the label of contents and question the storekeeper. If still in doubt, seek the advice of the Extension poultry specialist in your area. If the feed appears deficient in niacin, you can buy the vitamin in soluble form at most drug stores and add it to the ducklings' drinking water. A lack of niacin can stunt the growth of ducklings and cause them to have weak legs.

Speaking of legs, do not use newspapers or materials with smooth surfaces for bedding. A slippery surface underfoot can cause ducklings to suffer a condition known as spraddle-legs, which can cripple them.

Duck feed in the form of pellets or crumbles is better than a fine mash, which can cause them to choke. No matter what form the feed, always keep the water fountain full. But, to prevent waste, fill the trough only half way with feed.

Most guide books on the care of ducklings tell you to feed them a starter ration for the first two weeks and then switch to a growing ration. The problem here is that you usually have to buy the feed in 100-pound quanti-

An ideal setting for the small flock.

ties, and, in the first two weeks of their lives, your 15 little ducklings will only have consumed about 38 pounds. What do you do with the balance of 62 pounds? Don't discard it. Just keep it comin'. Keep on feeding it until the bag is empty. It will take the noisy little devils about four weeks to knock off 100 pounds of starter feed. Then switch to a grower-finisher ration, which will have a protein content of about 18 percent. Everything being relative, they will eat about 170 pounds of this before they average some 7 pounds each at seven weeks of age.

The feed-per-pound-of-grain ratio of White Pekins is about two and two-thirds to one (2⅔:1). That is, it takes about 2⅔ pounds of feed for the duckling to gain 1 pound, live weight. Ducklings eat more than chicks, but they also grow faster and larger. While a broiler-type chicken can weigh 4 pounds in eight weeks, a meat-type duck can grow to 7 pounds in seven weeks.

The growth of a duckling is most rapid and efficient during the first few weeks of its life. As an example, consuming less than 5 pounds of feed, the duckling can weigh about 3 pounds at three weeks of age. Yet, to gain the next 4 pounds that will raise them to a good slaughter weight of 7 pounds, it takes at least another 13 pounds of feed. The older and larger any bird or animal gets, the less efficiently it converts feed to meat.

Always store feed in tight containers with lids, so that rats, mice, squirrels and raccoons cannot get into it.

An Alternate Feeding Method

So far, we have been concentrating on getting the ducklings up to slaughter weight in the fastest time possible, by pushing them with a full-feed program, using high-protein commercial rations, and keeping them confined. There is another way to raise ducklings for meat.

For the first four weeks of their lives, care for the ducklings as previously described. Then, when they are feathered out, put them on range. Ducks are good foragers and will eat new grass, clover and green and leafy plants. They also eat grasshoppers, slugs, snails and most any small creatures that creep, crawl or fly. Some farmers place electric lights close to the ground in the duck yard, and they claim that when the lamps were turned on at night, their ducks thrived and grew fat on the insects attracted by the lights.

Continue to feed them a commercial growing ration, but cut the amount drastically. Instead of keeping feed in front of them all of the time, a once-a-day feeding toward sundown encourages them to forage during daylight hours. Provide plenty of drinking water, unless a small freshwater pond is available. A pool of stagnant water is a breeding ground for disease. And the ducklings must have shade.

Your 15 ducklings should have a yard space of at least 200 square feet, but this is a moot point and the figure arbitrary, because it all depends on the terrain and type of vegetation to be enclosed. Ducks raised in this fashion, on range, foraging for much of their feed in the form of plants and insects, are more lean at slaughter time. And you can cut feed costs by at least 25 percent.

Under this natural method of feeding, it will take more time, at least four weeks or longer, for the ducklings to gain good slaughter weight. You will have to have enough space for the yard, erect a fence, keep an eye out for predators and still put them under shelter during violent storms. And if your ducks are noisy, the neighbors might complain.

If you plan to turn your ducklings out on range to forage for part of their food, it might be better to choose a breed other than White Pekins, as they are bred for fast gains in confinement. The Rouen or Muscovy would be a good choice. Muscovies can fly fairly well when fully grown, but they should be roasting in your oven long before that time.

A fence from 2 to 3 feet high will keep the birds in since most domestic ducks, with the exception of Muscovies, are poor flyers. However, a fence from 4 to 5 feet high is needed to keep large predators out. A dog-proof fence should also have a strand of barbed wire at ground level and two strands at the top of the fence.

A roll of chicken wire of 2 inch diameter, 4 feet high and 75 feet long will cost about $18. Even buying chicken wire of 1 inch diameter will not solve the small predator problem, as rats and weasels can still get through. Chicken wire is flimsy stuff and, when erected outdoors, would need a great many posts set close together for any degree of strength. Sectional wire, of the type used for fencing hogs and sheep, is much sturdier. This wire is woven in rectangular patterns and comes in rolls 330 feet long, which cost about $100.

The problem here is that small predators can slip through this kind of fence easily, and the ducklings themselves can get out. Sectional wire is formed in a graduated pattern. The rectangular openings are smaller at the bottom and increase in size towards the top. By the time your ducklings are five or six weeks old and of good size, they should stay within the yard. Especially if you have provided them with feed, water and shade.

If you anticipate a predator problem, it might be best to pen your ducklings up in their secure house at night and turn them out in the morning. *Very important:* if you do put them inside overnight, and there is dry feed within their pen, they must also have plenty of water available so they can eat without choking.

Remember, if confined to a very small yard, the ducklings can decimate all vegetation and turn it into a mud hole. I would suggest fencing in the largest area that your finances and the available space allow.

Daily Care and Observation

Try to feed and water your ducklings on a regular basis. They are creatures of habit and the routine chore schedule should be consistent. After a few days, when the birds are eating out of the feed trough, remove the box tops that you used to start them off with. Keep feed in front of them all of the time, but don't fill the trough more than half full in order to reduce waste. If they don't clean up their food and it becomes mouldy, throw it away and replace with fresh feed. Try to keep their water fountain full at all times. Being waterfowl, ducklings are much sloppier in their natural feeding patterns than chickens. Especially on dry land. They can make the area adjacent to their feed trough and waterer very wet in short order. Check the condition of the litter near the waterer daily, and remove damp spots with a small dust pan and replace with dry bedding. Dave Holderread, in his excellent book, *Raising the Home Duck Flock*,* suggests that if you supplement your birds' diet with green plants or culled vegetables, cut them into small pieces and place in the waterer. This will help keep the food fresh and the ducklings will stay busy dabbling for goodies.

---

* Dave Holderread, *Raising the Home Duck Flock* (Charlotte, Vt.: Garden Way Publishing Co., 1978).

It is very important to observe your ducklings closely every day. If they are all huddled under the heat lamp, they need more warmth. Lower the infra-red lamp an inch or two. If they are crowding the cardboard guard, trying to escape the heat, raise the lamp. Try to reduce the temperature in their house by about 5 degrees a week, without making the ducklings uncomfortable. Contented ducklings will be moving freely within their pen, eating, drinking, and running from here to there for no apparent reason, but just full of the ebullience of youth.

Second Week. Remove wet and soiled spots, daily. Add fresh litter and stir the bedding. Don't let it become matted or caked. Decrease the floor temperature to 85° F. by raising the heat lamp. If the ducklings appear comfortable, enlarge the draft guard or remove it from the pen. Add fresh water to the fountain at least twice a day and keep the feed trough half full.

Third Week. Lower the temperature to 80° F. Follow a procedure of strict sanitation. Keep the waterer full and the bedding clean and dry. Remove any sickly or dead ducklings and bury or burn them.

Fourth Week. Reduce temperature to 75° F. Keep litter clean and dry. Add more bedding when necessary. About this time, the ducklings will have eaten their way through the original bag of starting ration and it's time to buy a bag of duck grower, which will be about 18 percent protein. Introduce the growing ration gradually, mixing it with the last of the starter feed.

Fifth Week. If your birds are to be confined indoors until they are ready for slaughter, add a second waterer and feed trough to their pen. Reduce the temperature to 70° F. They should be well feathered out, and, depending on weather conditions, you can turn them out on range at this time, provided you put them under shelter during heavy rainstorms.

Sixth Week. As you purchased the ducklings in late spring or early summer, they should be able to survive without artificial heat by this time, whether kept in or outdoors. If your ducklings are comfortable without the heat lamp, a 40-watt bulb will provide enough light for you to do your chores and for the birds to find their feed and water within the house. If they are out on range, they will regulate themselves by normal daylight hours.

Disease

Unlike baby chicks, ducklings are exceptionally hardy. Barring physical accident, it's possible to raise every one you buy to maturity. If you follow a strict program of good sanitation, with such a small flock there should be no disease problem. However, if more than one bird becomes droopy, listless, has no appetite or develops diarrhea, consult your Extension poultry specialist for advice and treatment.

Predators

If your ducklings are housed inside a snug pen for their entire life span of about seven weeks, no predators should bother them unless you let the family dog or cat in by mistake. However, rats and weasels can squeeze through a space 1 inch in diameter, and any sizable holes in the duck house should be covered over with strong wire mesh.

Rats will not normally attack a duckling after it's a few weeks old, but a weasel will take on even a mature duck, sometimes just sucking the blood or removing the head.

If your ducklings are out in a yard or on range by the time they are a month old, their most likely daytime foe will be a hawk. After dark, nocturnal predators like weasels, raccoons and foxes are out hunting. If you have reason to suspect the presence of these hungry duck fanciers, it might be best to herd your ducklings inside their house each night.

One rat can eat about 25 pounds of expensive feed and destroy three times that much in a year. A system of bait stations, using the anti-coagulant type of bait placed along runs and walls should control rats. Along with anti-coagulant bait, traps also help control rats, provided they are placed where family pets won't get caught in them. Also, don't give rats any places to hide or breed; get rid of trash piles and old lumber. Always store feed in tight containers.

After Seven Weeks

Your ducklings should be approaching slaughter weight. You have fed, watered and cared for them religiously, and now it's time for them to provide meat for your table. Pick out several of the largest birds and weigh them. When catching a duck, don't just grab wildly for legs or wings; they are easily injured. Instead, grasp the duck by the neck with one hand, and, with the other hand, hug it to you, enveloping the wings with your arms. After the bird has calmed down, you can support it with one hand under its body and the other gently holding its wings down. If the ducks only average 6 pounds or so, continue with your feeding program for another week. If they weigh 7 pounds, they are ready to be butchered. (Birds that have been out on range, foraging for part of their food, can take 12 weeks or more to reach optimum slaughter weight.)

One important rule for the first-time grower is don't make pets of your ducklings. When they are very young, everybody finds these downy, little, bright-eyed creatures irresistible. However, by the time they are seven weeks old and feathered out, particularly if they are an all-white breed like the Pekin, it's hard to distinguish between them, and the chance of forming a special bond with any one bird diminishes.

Preparation for Slaughter

You will need a sharp knife or axe, a length of cord, and a bucket or other container for catching the blood. If you want to make the process easier for yourself, you can buy a killing cone for about $4 (See p. 16). This is a funnel-shaped device that holds the bird securely, but exposes the head and neck to the knife.

Pick out two ducklings and isolate them in a crate or box the night before they are to be killed. Give them water but no feed so that the dressing-out process will be cleaner. If the isolation crate has slotted floors, manure and other waste can drop through and not soil the birds' feathers.

Slaughtering the Ducklings

One method is to hold the duck by the legs over a chopping block, restrain movement by placing its head between two spikes nailed into the block and cut off the head with an axe. Have a container directly below to catch the blood.

Another way is to put the duckling into a killing cone, and, as it hangs upside down, dispatch it with a knife. To do this, first grasp the bill or head of the bird with one hand and pull gently, but firmly, keeping a small amount of tension on the neck. With a very sharp knife, sever the jugular vein by cutting into its throat on the underside and slightly up and to the left just behind the head.

If you don't care to use an axe or buy a cone, tie the feet together with a short piece of cord and hang the bird upside down on nails driven into boards or rafters overhead. Restrain its head and neck as described before and proceed. When the bird stops struggling and all of the blood is drained out, cut off its head. However you kill it, it's best to have a container underneath the bird to catch the blood.

If possible, have an experienced person show you how to kill a duckling the first time.

The next step is to pluck the feathers.

Picking the Feathers

Fill a washtub or other large container with water heated to a temperature of 130 to 140° F. Hold the duckling by its feet, immerse it completely under water for about 1½ to 2 minutes, moving it up and down occasionally. The hot water helps loosen the feathers. A small amount of detergent can be added to the water to aid the process. A word of caution: if the water is too hot or the bird is immersed too long, the skin can become discolored and unattractive.

After the scalding, begin pulling out the feathers, but always pluck with the grain, that is, in the direction in which they are growing. Plucking against the grain can cause the skin to tear or break. If the feathers are hard to pull out, dunk the bird in the water for a few seconds more and try again. If pin feathers present a problem, they can be squeezed out, using pressure from your thumb and a dull knife.

When all of the feathers and down have been removed from the bird, a number of hairs will remain on the carcass. You can remove these by singeing them off. A small alcohol burner, candle, propane torch or the open flame of a gas stove can be used to singe these hairs. Be careful not to burn the skin or yourself.

Then, cut the feet off at the first joint. And now it's time to eviscerate.

Dressing the Duckling

Lay the bird on its back on a counter (put paper underneath to absorb blood and fluids), and, with a very sharp knife, make a very shallow cut about 3 inches long, in the soft area midway between the breastbone and the vent or rear end. The cut has to be shallow so that you don't pierce the intestines which lie just below the skin. A slight pull on the skin will give the knife a firmer surface for cutting.

After making the incision, pull any remaining layers of membrane apart with your fingers. Put your hand through the incision, reach up into the body, grasp the mass of viscera, which consists of the heart, liver, gizzard and intestines, and pull firmly until you have removed them from the body. One end of the large intestine will still be attached to the vent or anus. Make a circular cut around the vent, without penetrating the intestine, and remove entirely. Discard the intestines and put the heart into a pan of cold water. Attached to the liver, you will find a small green gland called the gall bladder. Remove this gland from the liver carefully; if it breaks, it will spoil any meat that it contacts. Put the liver in the pan containing the heart.

Reach up into the body cavity and feel for spongy tissue next to the ribs, and remove the lungs. They are soft and pink. Just above the tail on the back side of the bird is an oil gland. Ducks preen themselves with this gland. Make a deep cut around the gland and remove it.

Cut the neck off close to the body. With a strong pull, remove the windpipe and the esophagus (a tube leading from the throat to the stomach). Wash the carcass thoroughly, inside and out, with cold water.

Preparation of the Giblets

The heart and liver need only a good rinsing in cool water. To clean the gizzard, make a shallow cut around the outer edge, through the yellowish part, and peel it with your fingers. Discard the inside part which contains ground-up food. Wash the outer halves and place with the other giblets.

Chilling, Packaging and Freezing

It is very important that the carcass be cooled or chilled immediately after the bird is dressed. This helps prevent the development of harmful bacteria. The idea is to reduce the body heat of the bird down to a temperature of 34 to 38° F. as soon as possible. Even if you are going to eat the bird the next day, cooling and aging make the meat more tender. Ducklings to be frozen should always be chilled first. Twenty-four hours would be a reasonable time for cooling and aging. You can chill the bird in your refrigerator or soak it in ice water. If you plan on freezing the duckling and do use ice water for chilling, be sure to let it drain for at least 20 minutes before packaging. The less moisture in the package, the better the meat will keep.

When you prepare the duckling for freezing, remove as much air as possible from the wrapper after the bird is placed within. The tighter the plastic adheres to the skin, the better the quality of the meat will be. One way to create a vacuum is to insert a straw through the neck of the wrapper, draw on it to remove as much air as possible, twist the plastic tightly and tie securely.

Don't be disappointed if your first effort at dressing and packaging is less than perfect. The second bird will come out better. I would suggest that other members of the family participate in the whole operation, from start to finish. Then everyone can feel that they contributed something toward the ultimate roast-duck dinner.

Dressing Percentage of Ducks

Ducklings dress out to an average of 70 percent. The bird that weighed 7 pounds when alive will dress out to about 5 pounds, neck and giblets included.

Added Benefit—Fertilizer

Your 15 ducklings, especially if they have been closely confined within a building for seven or eight weeks, can provide you with a good supply of highly concentrated manure, litter included. Fresh poultry manure is not generally recommended for flower gardens or lawns, but it is an excellent additive for your compost pile.

References and Sources

Duck and Goose Raising, Publication 532, Ontario Department of Agriculture, Parliament Building, Toronto, Ontario, Canada, 1980. This booklet of 55 pages contains general information on the raising of ducks and geese.

Raising Poultry the Modern Way, Leonard S. Mercia, Garden Way Publishing Co., Charlotte, Vermont, 1975. Covers all aspects of raising poultry, including an excellent section on the production of waterfowl.

Raising the Home Duck Flock, Dave Holderread, Garden Way Publishing Co., Charlotte, Vermont, 1978. 178 pages of invaluable information by an experienced duck raiser who is obviously fond of his subject.

CHAPTER 4

GEESE

"The goose hangs high." Four short words. No excess verbiage. But what an image! When the bird is roasted, the toasty-brown, crackling skin on the outside tempts and teases with the promise of the tender, succulent feast contained within. If that image whets your appetite, consider the goose. Here is a way to produce 10- to 12-pound, oven-ready birds in 12 to 14 weeks.

This system is based on purchasing eight day-old goslings of a medium-sized breed, feeding them a commercial ration for 30 days, placing them out on range for at least two months and then butchering and dressing them yourself.

The eight goslings will cost about $32. Postal delivery charges can add another $5. In order for them to gain a live weight of 10 to 12 pounds in 12 to 14 weeks, you will feed the geese some 300 pounds of a commercial ration which will cost about $38. Necessary equipment such as feed troughs, waterers and a brooder lamp will cost another $20. At this point, your start-up costs for the small flock will be about $95. However, this system calls for the geese to be put out on range and forage for a large percentage of their food. Fencing can be costly, depending on the type of terrain and vegetation you are enclosing.

An avid vegetable gardener knows that carrots love tomatoes and slugs love beer. But how many know that geese love crabgrass? So, you could turn them loose on your lawn and let nature take its course. You would, of course, be careful where you walk, stepping gingerly in the dim light. Young geese, especially, are used for weeding strawberry plots, cotton and tobacco fields, orchards and vineyards.

If your garden is already fenced in, you could turn your goslings loose in it, provide them with shade and water, and your costs would be nil. You would take them out of the garden before any fruits or vegetables approach maturity. If you have to start from scratch, a standard roll of woven, sectional wire, 3 feet high and 330 feet long, costs about $100. This will keep the birds in and if there is anything tender and green growing within, it should suffice.

Geese are the most hardy and easiest of all poultry to raise. You can house them in a shed, garage, barn or any outbuilding. It is not necessary to have a pond or stream on your place to raise geese. They are quite agile when moving about on dry land. Although excellent swimmers, they are much less dependent on water than their close relatives, the duck and swan. They are larger than a duck, smaller than a swan and their legs are longer than those of either relation.

In the early stages of their life, they actually grow faster than the duck. A duckling on full feed can weigh 6 pounds in six weeks. A gosling the same age can reach 8 pounds during that same period. But, it takes the gosling at least three months to reach good size for slaughter. As with all animals and poultry, the older and larger the bird, the less efficient it is in converting feed to an increase in body weight.

You will spend about 15 minutes twice a day, every day of the week, feeding, watering and caring for your goslings. A capable child of 12 or more can share the chores with you. An adult goose can frighten a small child, or even a grown-up, with its hissing, flapping wings and threatening posture. However, your goslings will be on the dinner table or in the freezer before they reach that stage.

To sum up: eight day-old goslings; 300 pounds of feed; equipment like feeders and waterers; a suitable pen; a fenced-in yard or garden; and three months of light chores will provide you with enough delicious meat for all the holidays and a couple of extra special occasions thrown in.

Check Zoning Laws

Check with your town clerk in regard to local ordinances concerning the raising of poultry in your area. Young geese are not particularly noisy. They don't cackle or crow in the middle of the night. However, it would be a good idea to find out how your neighbors feel about your keeping geese.

Read About Geese

The supply of source books on raising geese is very limited. I would suggest that you contact the Extension Service poultry specialist in your area, every state has one associated with a college or university, and ask for literature on raising geese. I have always found these specialists to be very helpful and informative.

Visit Geese Growers

If people in your locality raise geese, visit them and ask about the care and feeding of goslings. They can tell you how they manage and perhaps direct you to a good source for birds. If they use terms with which you are unfamiliar, the following list may help:

Poultry. A general term covering all domesticated birds, including chickens, ducks, geese, turkeys, pigeons, peacocks and swans.

Fowl. A general term referring to chickens, turkeys, ducks and geese.

Waterfowl. A term designating birds that swim, i.e., ducks, geese and swans.

Goose. In general, a large, swimming bird related to ducks and swans. Strictly speaking, a female goose.

Gander. A male goose.

Gosling. A young goose, either male or female.

Green geese. Young geese that are full fed for their entire life span to promote rapid growth.

Housing

Before the goslings arrive, you will have to have a place to keep them. Their pen does not have to be fancy, but it has to be dry and free of drafts in order for the day-old birds to survive. You can use a shed, chicken house or any type of building that meets these requirements. If the building is large, such as a barn, you can partition off a corner for the goslings. The southeast corner is usually ideal. The floor can be of earth, wood or cement. The latter would be the easiest to keep clean and sanitary. Cement is also the coldest flooring, but you will use several inches of litter, like wood shavings, dry peat moss or chopped straw, to serve as bedding.

Eight day-old goslings don't take up much space, but by the time they are a month old, before they go out on range, they should have from 2 to 3 square feet of space, each. Now is a good time to prepare for their ultimate requirements. You will need space for the feed and water troughs and enough room so that you can move about easily within the pen, without stomping on any little goslings. If possible, give yourself a little extra room and make it easier to do the chores. It would not be reasonable to build a new house for the goslings. The cost would be prohibitive and you don't even know if you will like raising geese.

No matter what type of building you use for the birds, it has to be draft free. All windows and doors must close tightly. When they first arrive, you have to keep them warm enough so that they will survive. With such a small flock, ventilation should not be a problem. By the time they are larger and feathering out, they will be out on range with natural air conditioning.

This system is based on raising your first goslings in late spring or early summer in order to take advantage of mild weather. To start them off, you will use a brooder lamp to keep the goslings warm. If the building you house them in has no electricity, you can run an extension cord to their pen from the closest source of power.

I would suggest that you nail a fine-meshed wire on the inside of all windows so that sparrows and other wild birds cannot gain entrance and spread disease. The pen has to be secure against all predators, whether wild or domestic, including family pets. All doors must close tightly.

When to Buy Your Goslings

The best time of year for a novice to buy goslings is in late spring or early summer. Most hatcheries offer goslings for sale only from March through July. In the beginning, the main objective is survival, and temperate weather can only help in the success of your venture. Also, this can coincide with using them as weeders. For instance, if you purchase goslings in April and plant a garden in May, by the time the birds are about five weeks old and feathering out, they could be turned into your vegetable patch to do a job on weeds and grass. Obviously, this all depends on your climatic zone.

Geese are very hardy. It is not uncommon to raise every gosling to maturity. Always buy the best stock that you can from a reputable hatchery.

Where to Buy Your Goslings

The Extension poultry specialist in your area can help you choose a hatchery. Farm journals and rural newspapers are a good source for locating waterfowl hatcheries. The geese keepers you visited when you first became interested in geese can recommend a supplier. If you can buy goslings locally, you will save on delivery charges and you can always go back to the source for advice and help.

What Breed of Geese to Buy

In ancient Egypt, the goose was considered a sacred bird. The goose has always been popular in Asia and in Europe, from whence our present domestic birds came. However, there has been a steady decline in the use of geese as a source of food in the United States. Undoubtedly, certain prejudices have contributed to their lack of popularity. Goose takes a long time to cook? No more than turkey, pound for pound. Goose is greasy? It doesn't have to be. That's why they sell those racks that fit so neatly inside of roasting pans. Besides, before the day of hamburger helper, goose fat was the cook's best friend. While most all of the common breeds of geese are capable of effective meat production, the most popular breeds are:

Embden. This snow-white goose grows rapidly and matures early. Under expert management, the Embden gosling is capable of reaching a weight of 12 pounds in just 10 weeks. No wonder it's a favorite of commercial growers.

Pilgrim. You can tell the males from the females by the color of their feathers. The adult gander is all white and has blue eyes. The adult goose is gray and white and her eyes are hazel. This medium-sized breed is very good for meat production.

Toulouse. One of the largest breeds, the Toulouse is a handsome bird.

Toulouse geese.

The feathers on the back and wings are varying shades of gray, and the breast is white. A mature male can weigh up to 26 pounds.

White Chinese. With its long, swan-like neck and graceful manner, the White Chinese is a most attractive breed. They grow rapidly, mature early and are excellent weeders.

Buying Your Goslings

I would suggest that you buy eight day-old, as-hatched goslings of a medium-sized breed. As-hatched means that the goslings will be a mixture of male and female. Most waterfowl hatcheries do not offer a choice of sexed goslings, or, if they do, they charge at least $.50 per bird for their trouble. Although males grow a little faster than females, with just eight birds you won't save enough on feed costs to justify the added expense for specifying males when you order.

My own personal choice would be White Chinese goslings. As the old saying goes, "Every Goose a Swan," and the White Chinese rivals the swan in its gracefulness. On the more practical side, they mature quickly, are the best weeders and are less fatty than other breeds when dressed out for the table.

Preparation for the New Goslings

Before the goslings arrive, it is extremely important that you make their new home as clean as you can. Brush ceiling and walls and scrub the floor, if possible. If the building housed poultry of any kind in the past, use a disinfectant on the walls and floor. When using a disinfectant, follow the directions carefully so that the little birds won't be injured by chemical residue.

You will have to supply artificial heat for the goslings for the first four weeks of their lives. To do this, hang a 250-watt heat lamp in the center of their pen about 18 to 24 inches above floor level. The heat lamp should have a porcelain socket, a reflector and a metal guard over the bulb, in case it falls accidentally. An infra-red bulb is preferable to the white-light bulb. Infra-red light discourages cannibalism. Cannibalism occurs when an aggressive bird pecks at another one, pulling out feathers and eventually drawing blood. Other birds will then attack the victim and mutilate it or even kill it. Cannibalism is thought to be caused by over-heating, over-crowding, poor diet, lack of water or too much light. As you are only raising eight goslings and providing them with plenty of space, proper feed, an ample water supply and appropriate heat, cannibalism should not be a problem.

You will have to maintain a constant temperature of 85 to 90° F. in the goslings' pen for the first week. To do this, you will have to be able to adjust the height of the heat lamp, which you will hang overhead by a rope or chain. You will have to go through a trial and error process of raising and lowering the lamp to maintain a floor-level temperature of 85 to 90° F. To do this, hang a sturdy thermometer about 5 or 6 inches above the floor and at least 3 feet away from the lamp. As you experiment with the heat lamp, check the temperature of the thermometer until you find the correct height.

It would be wise to make a guard for the goslings. You can use corrugated cardboard boxes, taken apart and cut into lengths 1 foot high. Staple the ends of the cardboard together and erect the guard in a circle on the floor under the heat lamp and 3 to 4 feet away. The heat lamp should shine down in the center of the circular guard. The guard keeps the goslings close to the heat source, their feed and water, and also prevents drafts at floor level. The guard should be sturdy enough so that it can't fall over and injure the goslings. It is circular because goslings can panic when they are startled by loud noises or during a violent thunderstorm, and they have a tendency to flee in one direction. If the guard had square corners, they could all pile up and injure themselves.

When you have set the guard in place, put down a layer of sawdust, peat moss, wood shavings or finely chopped straw, from 3 to 4 inches high, on the floor to serve as bedding. This bed of litter has to be dry and absorbent to keep the goslings warm. When the litter becomes wet or dirty, remove the soiled part and replace with clean, dry bedding.

As the birds grow larger, enlarge the guard to give them more room, and add more bedding. After a couple of weeks, when the goslings have out-grown the need for a guard, you can remove it from the pen. Do not use newspapers or other smooth-surfaced materials for bedding. A slippery sur-face can cause the goslings to injure their legs and become crippled. Place the 1-gallon waterer and feed trough in close proximity to the heat lamp.

The day before the goslings are due to arrive, turn the heat lamp on and test the temperature at floor level, with the aid of the hanging thermome-ter. If the temperature is above 90° F., raise the lamp. If it's below 85° F., lower it. You will have to experiment in order to maintain a constant, cor-rect temperature.

You must have feed on hand before you bring the goslings home.

When the Goslings Arrive

If your goslings come through the mail, open the shipping crate in the presence of post office personnel and count the birds. Reputable hatcheries always insure parcel post shipments in the event of loss or damage en route. If there are injured or dead birds, file a claim with your postmaster and send it to the hatchery immediately.

As soon as you get the goslings home, take them out of the box carefully one by one, and dip their bills into the fountain, which should be filled with lukewarm water. The waterer should be deep enough so that they can sub-merge their entire bills. However, the fountain should not be wide enough for them to climb into, as they could become chilled or even drown. As the young birds use a great deal of water while feeding, you can put extra litter around the waterer. Or, if possible, place the fountain on a wire platform over a floor drain.

After you have introduced the goslings to water, place their bills into the feed trough, which should be equipped with a spinner bar on top that pre-vents them from climbing inside and soiling the feed. In the beginning, be-fore the goslings learn to eat from the trough, you can put a little bit of feed in small box tops and place these in various spots near the heat lamp. When the goslings are accustomed to eating from the trough, remove the box-top feeders. Always keep feed in front of the birds, but only fill the trough half full to reduce waste.

I have stressed the fact that you must maintain a temperature of 85 to 90° F. in the beginning, and close observation of the birds' actions will help you determine their degree of comfort. If they are huddled under the heat source, lower the lamp an inch or so. If they are crowding the edge of the cardboard guard and panting, raise the lamp. If the goslings are comfortable, they will move freely about their pen, eating, drinking and peeping contentedly.

Feeding the Goslings

Newly-hatched goslings can survive for up to 72 hours without food or water. This is because the yolk of the egg is absorbed into the body of the little bird shortly before it hatches. The yolk provides necessary food and liquid for the bird in the first hours of its life. Thus hatcheries are able to ship them for long distances. But, the sooner you feed and water them after they arrive, the better off they will be.

In the beginning, the goslings should be fed a goose-starter ration of about 22 percent protein, in crumble or pellet form. If you cannot buy goose feed at your local store, duck-starter will do just fine. If that's not available, you can use a non-medicated chick- or turkey-starter for your goslings. Medicated chick-starter can contain drugs that are harmful to waterfowl. If in doubt, ask the Extension poultry specialist in your area for advice.

Feed in the form of crumbles or pellets is better than a fine mash, which can cause the birds to choke. If nothing else is available, you can feed them mash. Remember, no matter what form the feed comes in, always keep the water fountain full. You will start out by feeding the goslings free-choice; that is, keep feed in front of them at all times. But, to prevent waste, only fill the trough about half way.

Goslings can be turned outside to begin eating weeds and tender grasses when they are two weeks old. But, at that age, they won't have enough feathers on their back to protect them from chilling rains. It would be better to keep them confined or semi-confined, until they are four weeks old and well feathered out. Everything being relative, at four weeks of age they can average about 5 pounds in weight each. At this point, they will have eaten about 8 pounds of starter ration each, or a total of about 64 pounds for the eight goslings.

When the goslings are a month old, you can switch them to a goose- or duck-grower ration of from 15 to 18 percent protein. Poultry feeds usually come in 100-pound bags, and you will probably have a considerable amount of starter left over, which you can mix with the grower feed. The growing ration is cheaper and at this stage of their growth, the goslings don't need the higher-protein feed.

Geese on Range

At four or five weeks of age, the young geese will have grown their protective back feathers and can go out on pasture or into the garden as weeders. If on pasture, they should be confined to an area they can keep clipped down, as they relish short, tender grass over the tall, dried-out, fibrous kind.

If you plan to have them in a vegetable garden, put them in before the weeds get too much of a head start. A word of caution: don't put them in a pasture or garden that has been sprayed with poisonous chemicals.

Geese eat a variety of weeds, and six to eight goslings can effectively weed up to an acre of strawberries. Some commercial growers purposely keep their geese hungry by cutting down on the store-bought feed, so that the goslings will be better weeders. For our purpose, I would still offer the birds a free choice of pellets, as they will grow faster. However, I would place their feed trough at the far end of their pasture or range, forcing them to walk past tender grass and tempting weeds on their way to the pellets. This is sneaky, but it works.

While it is possible to keep the goslings completely confined within a house or pen until they reach slaughter weight, I don't recommend it. The geese on range will be healthier and less fatty than those raised in total confinement. Also, by the time they reach a weight of from 10 to 12 pounds, they will eat about 20 pounds less of costly feed, per bird, if allowed access to forage. With eight birds, this can mean a savings of at least $20 in feed costs.

You can also feed whole grains such as oats, wheat, corn or barley to the goslings when they are about six weeks old. These grains can be used to supplement the pellets or feed. Shop around. It all depends on where you live. In my area, a 100-pound sack of corn costs $10, while a bag of goose-

grower ration costs $12.50. This is a decided savings. However, feed stores adjust their prices according to locality and if you live close to an urban area, you may find that the difference in cost between commercial feed and whole grains is negligible.

The feed-conversion ratio of a gosling is most efficient during the first few weeks of its life. For example, at nine weeks of age, a gosling placed on range and provided supplemental feed can weigh about 10 pounds. The bird will have consumed about 20 pounds of pellets, plus much green forage by this time. This is a feed-conversion ratio of two to one; that is, 2 pounds of feed for 1 pound of weight gain. In order to add 2 more pounds to that same gosling's weight and bring it up to a good slaughter weight of 12 pounds, you will have to feed it another 10 pounds of pellets. As you can see, the bird's efficiency in converting feed to gain-in-weight decreases drastically as it grows in size.

Always store feed in closed containers, so that mice, rats, squirrels and raccoons cannot get into it.

When your goslings are on range, whether in a pasture or vegetable garden, they must have plenty of water and sufficient shade. You can distribute shallow buckets or pans in various places within the enclosure to provide water. Don't use a receptacle that a young bird might crawl into, or get hung up in, and possibly drown.

Trees can provide natural shade, or, lacking those, you can build temporary shelters with small pieces of plywood and old boards. No sides are necessary, just a slanted plywood roof affixed a couple of feet off the ground. Another alternative, small "A" frames, also work fine.

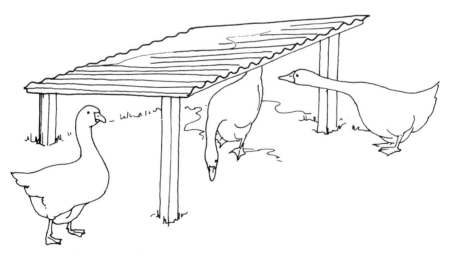

Temporary shelters, built from scrap lumber and any other suitable materials, may be used for geese on range.

A fence from 2 to 3 feet high is enough to keep the goslings in; domestic geese are poor fliers. An adult goose needs a sturdy tail wind and about 20 running feet just to get off the ground. A dog-proof fence should be from 4 to 5 feet high with strands of barbed wire at top and bottom. However, dogs are more likely to chase after baby chicks and ducklings that will run away from them. Geese are more aggressive and even a half-grown gosling, especially a male, will often turn to face its foe, hunker down close to the ground, thrash its wings, stretch out its long, sinuous neck and hiss threateningly all the while. In any case, it's up to you to keep family pets under control.

Although it's possible for the young goslings to climb through woven wire, if you have provided them with feed, water and shade, they should stay within the enclosure.

Daily Care and Observation

Set up a regular schedule for feeding and watering the birds, morning and night. Observation is the key to good management. You have to do more than just dump feed in the trough and fill the waterer twice a day. Observe the birds. Do they appear too hot? Too cold? Are they comfortable and content? Are they eating and drinking? Is there a droopy-looking gosling standing off by itself? After a short time, you will automatically check your birds every day.

When the birds are eating out of the trough, remove the box tops you started them with. If their feed becomes mouldy, throw it away and replace with fresh pellets. If the litter next to the water fountain is damp and foul smelling, remove it and add fresh bedding. Mouldy litter can be a breeding ground for disease.

Second Week. Keep feed trough half full. Keep water fountain full. Check condition of litter adjacent to waterer. Decrease the temperature of the pen to 80° F. by raising the heat lamp an inch or so. You will have to experiment. Observing the actions of the little birds is a better clue to their comfort than a thermometer. If the goslings appear comfortable, you can enlarge the circumference of the draft guard by adding more cardboard sides, or remove it entirely. Remove any obviously sick birds and dispose of them.

Third Week. Observe the rules of strict sanitation. Keep bedding clean and dry. Provide fresh water and feed. Lower the temperature at floor level to about 75° F.

Fourth Week. Stir litter and make sure it is clean and dry. Reduce temperature in pen to 70° F. If the days are warm and sunny and the temperature outside is about the same as inside, you can discontinue the heat lamp. If nights are cool, use the lamp at night. By this time, the goslings should have a good growth of back feathers. Set up your fencing arrangement in preparation for putting them out on range.

Fifth Week. Buy a bag of growing ration of from 15 to 18 percent protein. Mix the grower with the remainder of the starter feed.

Weather permitting, turn goslings out on pasture or into the vegetable garden. Provide feed, water and shade outdoors. Observe them closely. If you feel it necessary, either for reasons of inclement weather or a possible predator problem, pen them in at night. If you put them in at night and there is dry feed in their house, make sure they have water to wash it down with.

Sixth Week. The goslings should be weeding with a vengeance by this time, and eating the short grasses. Supply water and pellets free-choice. Place feed trough at far end of fenced enclosure.

Seventh to Ninth Weeks. Check fences. Supply ample water and feed. If pasture is deteriorating, increase pellet ration. When the birds are about nine weeks old, catch a couple of the biggest goslings and weigh them. To catch a goose, grasp its neck with one hand and with the other hand, hug the bird to you, enveloping its wings with your arm. Don't grab it by the legs or feathers, as they are easily injured. You can use a pan or hanging scale for weighing. If the gosling weighs close to 10 pounds, you are right on target.

10th to 12th Weeks. The goslings should be nearing good slaughter weight. Check condition of pasture, and feed the geese accordingly. Provide a constant supply of water. To prevent sunstroke, make sure the birds have shade. If they have been enclosed in a vegetable garden, take them out before any fruit begins to ripen. This may necessitate fencing in another yard for them. This yard may be quite small as the birds have now reached the finishing stage, and a full ration of pellets will help them reach optimum weight. Weeds and grass are not sufficient for finishing the gosling. Catch and weigh a couple of birds again. If they weigh about 12 pounds, you have done a masterful job. If they weigh less, continue feeding them free-choice for another week or so, until they gain the desired weight.

Disease

I find that geese are the most hardy of all poultry. The chances are good that you will raise every gosling to slaughter weight, barring physical accident. If you follow a program of good sanitation when they are penned up for the first four weeks of their lives, there should be no disease problem. If, however, more than one bird suffers a loss of appetite, becomes droopy or develops a discharge or diarrhea, consult your Extension poultry specialist for advice and treatment.

Predators

As your goslings will be housed inside a secure pen for the first month of their lives, unless the family dog or cat gets in by mistake, there should be no predator problem. Rats and weasels, however, can squeeze through a space 1 inch in diameter, and any sizable holes in the goose pen should be covered over with strong wire mesh. Usually, rats will not bother a gosling after it's a few weeks old. However, one rat can eat or destroy up to 100 pounds of feed within a year. You can't afford to support rats. To begin with, get rid of any trash heaps or old lumber piles and don't give them any places to breed or hide. Anti-coagulant type bait, plus traps, is recommended for rat control.

By the time the goslings are four weeks old and out on range, they are a bit too big and bodacious for most hawks to bother with. So much for daytime foes. After dark, predators like weasels and foxes are out a-hunting. I've lost plenty of chickens and ducks to the blood-thirsty weasel, but never a goose. And none of my goslings ever greased a fox's chin. However, there's always a first time. If you suspect that nocturnal predators may cause a problem, you can pen your little flock inside their house at night.

After 12 Weeks

If the goslings weigh about 12 pounds, they are ready for slaughter. If they are only 10 or 11 pounds, continue feeding them for another week or two.

Embden geese (background), Rouen and Pekin ducks.

Do not make pets of your goslings. It's very hard to butcher good old goosey-gander. You have fed, watered and cared for these birds for at least three months, and now it's their turn to provide a succulent feast for your table.

Killing, Plucking, Evisceration and Preparation for Freezing

If possible, it would be best if you could have an experienced person go through the whole process with you, step by step, from start to finish. The killing, plucking, dressing and preparation for freezing of a goose is not that much different from doing up a duck. Except now you are dealing with a much larger bird. The gosling has more and bigger feathers, much more down and is harder to pluck. The temperature of the water to be used for scalding a gosling before plucking should be from 145° F. to 155° F., a bit hotter than for ducks. Due to its size, the gosling should be immersed for up to 3 minutes. Eviscerate the bird and chill and package it in the same manner as practiced with ducks. Obviously, the larger carcass requires a larger freezer bag.

Dressing Percentage of Geese

Although the figure can range from 68 to 73 percent, geese dress out to an average of 70 percent. The 12-pound gosling will dress out to about 8½ pounds, neck and giblets included.

About Feathers

Goose feathers are valuable and can be used for making coverlets or other bedding. To wash them, use lukewarm water and add a little detergent, borax or washing soda. Rinse the feathers, wring them gently, and then spread out to dry. Generally, it takes the feathers from three geese to produce 1 pound of feathers. The bedding and clothing industries will purchase goose feathers if you have a substantial quantity of them.

Fertilizer

Your eight goslings, having been confined within a house for four weeks, will provide a small supply of highly-concentrated manure, litter included. This would be an excellent additive for your compost pile. In addition, while they have been weeding your pasture or garden, they have been spreading a high-quality fertilizer on the soil.

References and Sources

Raising Poultry the Modern Way, Leonard S. Mercia, Garden Way Publishing Co., Charlotte, Vermont, 1975. Contains an excellent section on waterfowl production.

Duck and Goose Raising, Publication 532, H.L. Orr, Ontario Department of Agriculture, Toronto, Ontario, Canada, 1980. This 55-page booklet contains general information on the raising of ducks and geese, from hatching eggs to eventual evisceration.

Backyard Poultry, Rte. 1, Waterloo, Wisconsin. A monthly magazine directed toward the small-flock owner that features articles on most types of poultry, including geese.

CHAPTER 5

RABBITS

There is a female rabbit in a cage in my backyard and every time I look at her, her fabulous potential never ceases to amaze me. This little 10-pound doe is capable of producing her own weight twelvefold—in edible meat—within one year. And that's the best reason I know for raising rabbits.

She lies there, stretched out lithe and lean, 10 pounds of soft, white fur. And, although she's safely enclosed in a wire cage, the pink eyes are always searching and the pink nose and busy white whiskers, testing the wind for scent of friend or foe.

It seems incredible that this little doe, under expert management and optimum conditions, can bear eight litters of eight babies each within one year. That's 64 little rabbits, which, within that same year, can grow to fryer weight, or at least 4 pounds each, for a gross live weight total of more than 250 pounds. After butchering, her offspring can dress out to more than 150 pounds.

While it would not be reasonable for a beginner to expect to duplicate the high production figures of commercial growers, it is quite probable that you could raise four litters of eight babies each from one female rabbit within a year. This would give you a total of 32 young rabbits, which, as fryers, could dress out to about 64 pounds of meat. Here is a system for producing two months of Sundays' worth of delicious dinners.

This method is based on purchasing a trio, that is, two does (female rabbits) and one buck (male) of a medium-sized breed. You would then buy or build your own cages, manage the breeding process and feed the young bunnies a commercial growing ration for about eight weeks. At that time, you would then butcher and dress the fryers out yourself. You should own or have access to a large freezer. As you will begin your new venture in temperate weather, late spring or early summer, you can house your rabbits in a shed, garage or partitioned-off barn. You can even raise them outdoors. At this time, it would not be practical to build a new structure to house them.

You will spend about 15 minutes twice a day, every day, in the feeding, watering and observing of your rabbits; about 3½ hours a week. The rabbits must be cared for every day, seven days a week, holidays included. However, after some instruction, any responsible child of 12 or more can tend the rabbits. One of the most popular projects for children in 4-H clubs is raising rabbits. It teaches responsibility.

The trio of rabbits will cost about $45. If you build your own cage, the material and tools will cost about $23. The waterer and feed trough will cost another $7. Eight little bunnies will eat about 128 pounds of feed in the eight weeks it takes for them to grow to slaughter weight. The feed will cost about $19. It will cost about $10 to feed one female rabbit through the gestation period and up to weaning time. Your start-up costs will be about $104.

Keep in mind, the cage and troughs can be used again and again, and the doe rabbit will bear many more litters. Also, I am including the price of the second doe in the start-up costs, and if she bears a litter of eight, you can net another 20 pounds of meat just for the cost of feeding her and eight bunnies.

Check Zoning Laws

Before even considering the raising of any animals in your backyard, check first with local officials in regard to the zoning ordinances of your village or town. There is little point in becoming enthusiastic about keeping rabbits if it is against the law. As I write this, a man who still lives on the farm he was born on in Orange County, New York, is being fined $25 a day for harboring one rooster, two hens and three rabbits, in violation of a town ordinance.

Given his day in court, this man offered that the rabbits were his children's pets, but his plea cut no ice with the village justice, who claimed that rabbits were not considered domestic animals. "In this town," the judge decreed, "you can keep three dogs or six cats, but no rabbits."

No matter that man has been raising rabbits, especially for food, for more than 200 years. No matter that rabbits produce useful products like meat and fur pelts, and are important in medical research.

First things first, check out the law before you build that cage.

When to Buy

Domestic rabbits will breed and bear young the year around and bear little resemblance to their wild brothers. The common cottontail can have two or three small litters in late spring or early summer. If there are any lingering traces of atavism, certainly then the most opportune time for a beginner to start raising rabbits would be in the spring or early summer, taking advantage of temperate weather.

Talk Rabbits

I would suggest that you visit every rabbitry that you can find within a reasonable distance from your home.

I have never met a rabbit keeper who wasn't proud to show his or her animals to me and "talk rabbits." You can pick up all kinds of helpful information in this way. Rabbit raisers will tell you where they purchased their rabbits, what they feed them, how much it costs, where to buy or how to build cages, and they can give you tips on breeding and caring for the young.

What Breed of Rabbit to Buy

There are at least 50 breeds of domestic rabbits, ranging in size from a tiny 2-pound Polish dwarf to a 20-pound Flemish Giant. Some rabbits are called Satins, and, you guessed it, they have satiny fur. Angoras are raised for their long wool, and Chinchillas° for fashionable winter coats. Some breeds are particularly good for medical laboratory purposes and some especially for meat. As the title of this book suggests, we will be mostly concerned with those breeds that are best for efficient meat production.

Among the most popular breeds in the meat category are the New Zealand White, Californian, Champagne d'Argent and Satin. These are all termed medium-sized rabbits, weighing from 8 to 12 pounds at maturity. There is good reason for raising the medium size for meat. Although the very small breeds take up less space and eat less feed, they have to be fed and carried longer than eight weeks to net a reasonable amount of meat. The very large, giant types, weighing from 13 pounds on up, have larger appetites and take much longer to grow and are not of favorable size for slaughter at eight weeks. Butchering an eight-week-old giant would be like picking a green tomato when there's still plenty of hot summer sun left to ripen the fruit.

The medium breeds, such as New Zealand Whites and Californians, being the most popular in the United States, have been raised by the hundreds of thousands, for research in management, nutrition and health by the major feed milling companies like Purina, Carnation, Agway and Maritime Co-op. The New Zealand Whites and Californians have been bred to perform with a high standard of predictable consistency. Time after time and litter after litter, the white does will bear at least eight healthy young, and the babies will gain an average weight of from 4 to 5 pounds within eight weeks. This is fryer weight. Unlike many Europeans, who prefer an older and bigger rabbit to roast for their dinner, Americans have a taste for young and tender fryers.

I would suggest that, as a beginner, you start out with either New Zealand Whites or Californians. First, as they are the most popular, you will have a better selection to choose your rabbits from. Second, as more research has been done with them, in a manner of speaking, they are the most improved breeds.

If, however, in your travels around the countryside (and I hope you

---

° Not to be confused with the chinchilla, a small rodent having a valuable coat of fur.

would spend a great deal of time looking at and "talking" rabbits before you buy any), you find a particular breed or color that takes your fancy, and you want to own red, black or blue rabbits, then by all means get them. The meat will taste just as good. And you'll be happier working with them. Whether we are discussing beef cattle or rabbits, there is no *one best breed*, but only outstanding individuals or strains (family bloodlines) within each breed.

I would recommend purchasing a trio; that is, two does and one buck, for starters; preferably, one junior doe (younger than six months) and a senior doe (over six months and having had a litter or two already). Does normally reproduce for three to four years. Some exceptional individuals reproduce for six years. The buck can be either a junior or a senior. The younger one will give you more years of service. But, an older buck will have proven his worth as a sire.

Bob Bennett, in his excellent book, *Raising Rabbits the Modern Way,** recommends starting out with four rabbits, two does and two bucks. And he backs up his case with careful thought and many years of experience. But, it is my personal feeling that for a raw beginner interested only in raising meat for the table, the second buck is extra baggage. One more mouth to feed. An extra cage to maintain.

One buck can service at least 10 does and, under special conditions, up to 20. If, after a year or so, you decide that you are really hooked on raising rabbits, then there is plenty of time to buy another buck. Even with just two does, by the end of the first year and under minimal management conditions, you will still have produced 50 or 60 young rabbits. And if you have changed your mind about rabbits, you will not be in too deep.

When you buy your first rabbits, get the best stock that you can find and afford. Cross-bred junior does will cost about $5, older does around $10, and the bucks about the same. Crossbreds are a mixed bag and you don't know how the offspring will turn out. Purebreds will cost from $15 to $25 apiece, but are worth it. And purebreds have papers. Honest. Even the rabbit, with its acclaimed regenerative powers, capable of producing up to eight litters a year, has a pedigree. Is the pedigree worth the paper it's written on? Yes. It tells you that the rabbit raiser took the time and trouble to keep records. Pedigrees attest to a background of consistent production. That's most important: consistency! That's what puts a pro football team in the Super Bowl, year after year. And that's what puts scrumptious meat on your serving platter, month after month.

Where to Buy Your Rabbits

Now that you've decided on a breed of meat-type rabbit and how many you are going to buy, it's time to decide *who* you will buy from. At this

---

* Bob Bennett, *Raising Rabbits the Modern Way* (Charlotte, Vermont: Garden Way Publishing Co., 1975).

California rabbits.

stage, you've checked out every rabbitry from the small backyard hobbyist to the large, commercial operation. I would suggest that you subscribe to *Rabbits* magazine (address at end of chapter). I would write to the breeders who advertise and ask their prices for breeding stock. Keep in mind that, as a customer, you are usually responsible for the shipping charges.

Go to every county fair that you can. There are usually rabbits on display. Attend every rabbit show in your area. *Rabbits* magazine lists current shows from coast to coast. The Extension office of your county agricultural department can give you information about local rabbit breeders. The raising of rabbits is very popular with 4-H club members, and the kids may have good stock for sale. Your local feed store can give you the names of people in your area who raise rabbits.

Buying a Rabbit Cage

A standard rabbit cage made of wire and with dimensions of 36 x 30 x 18 inches, already set up and ready for use, costs $28 at the farm supply store in my area. This price includes a door but no feeders or waterers.

The mail-order equipment dealers list standard wire cages for an average of $18. These are shipped flat and you have to assemble them upon arrival. They are sent via parcel post or United Parcel Service, and you pay the shipping charges, which can amount to about $5 per cage. Or you can build your own.

How to Build a Standard Wire Cage

NECESSARY MATERIAL AND PARTS

- Eleven feet of 16-gauge, welded, galvanized wire with a mesh of 1 inch by 2 inches. This piece is 18 inches wide and is for the sides.
- One piece, 36 x 30 inches, of 16-gauge, welded, galvanized wire with a mesh of 1 x 2 inches. For the top of the cage.
- One piece, 36 x 30 inches, of 14- or 16-gauge, welded, galvanized, wire with a mesh of ½ inch x 1 inch. This sturdier wire is for the cage floor.
- One piece, 14 x 14 inches, of 16-gauge, welded, galvanized wire with a mesh of 1 x 2 inches. This piece is for the door of the cage. You are going to have to cut an opening on one side of the cage, at least 12 x 12 inches, large enough for easy access for cleaning and removing rabbits, and placing the nesting box. The door you fashion should overlap on all sides, thus the size of the door depends on the opening you cut.
- One pound of J clips, for fastening the wire together.
- One pliers for use with J clips.

Lay the 11-foot long, 18-inch wide, wire piece flat on a smooth floor. Measure 36 inches and with a rubber mallet, bend the wire around a straight-edged wooden board, forming one corner. Measure 30 inches and bend again. Measure 36 inches and bend. Fasten the loose ends together with the J clips spaced every 2 or 3 inches apart. You now have a rectangle, with dimensions of 36 x 30 inches that will serve as the sides of the cage.

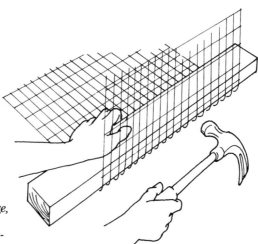

When building your wire cage, form corners by bending the wire pieces around a straight-edged wooden board.

Cut a piece of wire 36 x 30 inches, of the 1- x 2-inch mesh, and fasten to the sides. This is the top of the cage.

Cut a piece from the heavier 14-gauge wire with the ½- x 1-inch mesh, to the dimensions of 36 x 30 inches. This will be the floor. Before fastening, determine the smoothest side of this piece. In other words, if there are sharp edges on one side of this close-meshed wire due to the galvanizing process, fasten the smoother side *up* so that the rabbit's feet will not become injured.

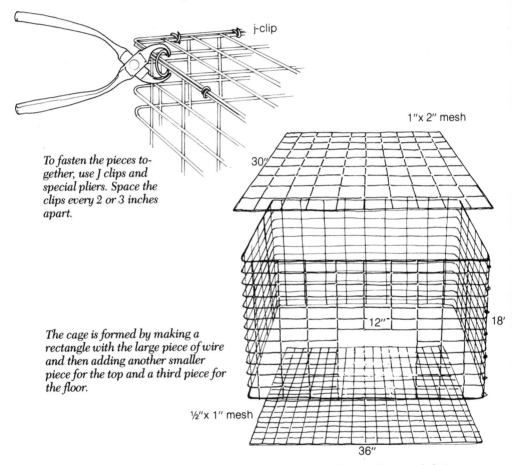

j-clip

To fasten the pieces together, use J clips and special pliers. Space the clips every 2 or 3 inches apart.

1"x 2" mesh

30"

The cage is formed by making a rectangle with the large piece of wire and then adding another smaller piece for the top and a third piece for the floor.

12"

18'

½"x 1" mesh

36"

You now have a wire hutch, but no way to get inside it. What to do? Cut an opening, 12 x 12 inches, in the widest (36-inch) side. Cut a door, 14 x 14 inches, to fit and place inside the cage, hanging it from the top of the cage with wire or hinges, so that it overlaps the opening on all sides. How do you get a 14-inch door through a 12-inch opening? Diagonally.

You can fasten a small wire hook on the top of the cage, about 12 inches in from the edge, so that if you want the door to stay open temporarily, for cleaning, etc., just hook the wire to the door.

If you decide to buy a trio of rabbits, two does and a buck, you will need three cages right off. Within a couple of months, after the does have their young, if you decide to wean the babies from their mothers, you will need another cage. It seems that you can never have enough cages.

At my local hardware store, the rabbit wire costs about $12 for a standard-sized cage. One pound of J clips is $2 and the special pliers retail at $9. This adds up to $23 and your time. The pliers can be used to make 100 cages; thus, their cost cannot be charged against just one cage.

When you finally make your decision, you are banking on the honesty and reputation of the seller. If you buy purebred rabbits, the chances are that the seller will stand behind them.

What to Look For

If you don't purchase your initial breeding stock through mail order, but buy them locally instead, look for alert, bright-eyed, good-sized animals. The does should be long and lean with shiny coats, not woolly, and have at least eight teats (nipples). There should be no visible defects like weak fore or hind quarters or buck teeth. A friend of mine has a doe who is excellent in every way except that she has two lower front teeth that keep growing out to a length of two inches or more and interfere with her ability to chew. Therefore, the teeth have to be cut back periodically.

The buck you choose should have the same qualities as the does: bright-eyed, sturdy, vigorous, with a good coat of fur. He will be blockier and shorter than the female, stout but not too fat. You should also examine him to see that his sexual organs, the testicles, are well developed.

How to Pick Up a Rabbit

In order to examine your rabbits, you will have to pick them up. You might as well take the plunge right off and get used to it. Gently but firmly, grasp the animal by the loose skin at the nape of the neck and shoulders with one hand and slide the other hand around the rump, hugging the rabbit to your waist. Like carrying a football through scrimmage. Although former president Jimmy Carter was allegedly attacked by a swimming rabbit, I have never been, nor do I know anyone who has ever been, bitten by a rabbit. But, I have been severely scratched by the sharp toe nails on the thrashing hind feet of a struggling animal. Stay calm, be cool, and they will take their cue from you. Never pick them up by their ears.

The Best Rabbit Cage

Rabbits take up little space. You can keep them in your backyard shed, garage or an old chicken house. But, before bringing them home, you have to have a place to put them. The keeping of rabbits does not automatically result in bad odors, rats and bugs as some people may think. Rabbits will be just as clean as you allow them to be. And that's where the type of cage you use comes in.

The best kind of cage (hutch) for rabbits is made of wire mesh. The most important requirements for a hutch are ventilation and sanitation and the all-wire cage provides these needs much better than one made of other materials such as wood or hard board.

The standard cage for medium-sized rabbits is 36 inches wide, 30 inches deep and 18 inches high. This hutch allows the animal to move around a bit and stretch, but takes up no more room than is necessary, and, when you are raising rabbits, space becomes an important factor, as they tend to expand. It is big enough for one doe to raise a litter of eight young up to the age of eight weeks without being too cramped or confined. Domestic rabbits don't have to hop around like their wild forebears. In fact, the quieter they are, the quicker they will grow and gain, and that's the whole idea.

The all-wire cage provides ventilation on all sides and is easy to clean. Urine and droppings go through the mesh floor. Fecal matter will collect on the sides and bottom of a wooden hutch, and this can lead to foul odors and possible health problems. It's much easier to keep your rabbits clean and prevent diseases than it is to cure sick animals.

Living on wire is not harmful to rabbits as the floor of a wire cage is usually constructed of a narrow ½-inch by 1-inch pattern. In the usual sitting position of the rabbit, the heavily-furred hock (hind foot) supports much of the animal's weight.

Housing Your Rabbits

You can use a shed, garage or other outbuilding to house your rabbits. If you use a building, make sure the windows are functional and the door is secure. You will have to adjust the windows to prevent drafts in cold weather and provide ventilation in extreme heat. The door will keep predators out. Hang the cages from rafters, with sturdy wire, at a level comfortable for you to work with, about 3 to 4 feet high. If the building you use has a cement floor, you will need a metal pan to catch the animal droppings. If you want to be really clever, you can train your rabbits to do their thing in a designated place by moving their droppings under a specific area of the cage for a week or so. They will tend to seek that area out, by scent, when nature calls.

If your rabbits are housed indoors, but over a dirt floor, you can add lime, hay and superphosphate to the heap, as Bob Bennett recommends in his book, *Raising Rabbits the Modern Way*. This requires that you remove the manure only two or three times a year.

If your rabbits are to be kept outdoors, you will have to protect them from the hot sun, rain, snow and cold winds. Suspend the cages above the ground at a comfortable height for working and handling. I would suggest that they be over a fairly well-drained area and arranged under the long branches of a hardwood tree, if possible. This will provide some shade in summer and a bit of protection from winter winds. Do not obstruct the cage floors so that the droppings cannot fall through. Add superphosphate and lime to the manure pile from time to time.

You would also need to provide plywood boards or heavy canvas for the top and sides of the cage in the dead of winter. Rabbits cannot take intense heat or cold winds. They must be protected from these extremes. On the other hand, they *can* take the cold. They will grow a fabulous winter coat to combat sub-zero temperatures.

Also, if your rabbits are housed outside, you should erect a protective fence around the hutches, at least 10 feet from the hutches. The fence can be made of boards, 4 or 5 feet high, or from sectional wire, with a strand of barbed wire at the very bottom, on the outside, and one or two strands of barbed wire at the top. You are not only protecting the rabbits from physical injury, but, just as important, stress. Stray dogs are the worst offenders.

Now that you've bought your rabbits, brought them home and installed them in hutches, you've got to feed them.

What to Feed Your Rabbits

The best feeds for your new rabbits are the complete rations in the form of pellets manufactured and packaged by the major milling companies like Purina, Carnation, Agway and Maritime Co-op. After many years of nutritional research, these companies have put together a well-balanced ration of protein, energy, vitamins and minerals in palatable form.

You can have the best of livestock, but without the proper feed, they won't produce up to their capability. Even if you *could* grow your own cereal grains and come up with the right formula of amino acids, sugar, starch, fat, crude fiber and add vitamins and minerals, by the time you put it all together in substantial quantities in a handy form, it would be more trouble than it's worth. If you had to buy the individual ingredients to mix up a batch, the cost would be prohibitive.

The feed companies caution, and rightly so, that if you add oats, barley or other grains to their pellets, you can dilute their carefully planned premixed nutrients and thus nullify the benefits of their complete rations.

You *could* feed your rabbits on hay, greens and roots, and they would probably survive, but the does would not produce strong, healthy litters time after time; they would not have the milk and the young would not grow as fast, and that's the main idea.

As a rule of thumb, do not feed greens or vegetables to any rabbit under the age of six months. This can result in serious intestinal problems in the young. You can give older rabbits a bit of lettuce, carrot or other roots as a special treat on occasion.

In regard to salt, the pellets contain a small quantity, but if you feel more is necessary, salt spools are available at farm supply stores for about $.25 each.

I would strongly recommend that you keep all feed sacks in a tightly enclosed metal container to prevent spoilage and waste by rodents.

You cannot go wrong by feeding just the pre-packaged rabbit pellets, which cost $15 for 100 pounds in my area, and by supplying plenty of fresh, clean water.

How to Feed Your Rabbits

The dry (non-pregnant) doe should receive 4 to 6 ounces of feed per day. Adjust amount of feed to weight. This means observe her closely and if she seems to be getting overly fat, decrease her ration. If she seems to be too thin, increase her feed.

The pregnant doe should be fed from 4 to 8 ounces per day. Again, adjust to weight and appetite. If the doe weighs about 10 pounds, and you start her on 6 ounces, and she cleans it up quickly and seems hungry for more, then add a couple more ounces.

The lactating (nursing) doe should be fed free-choice.

When the doe kindles (gives birth), cut her ration in half to prevent too much milk. After a couple of days, increase her feed gradually, at about ½ ounce per day, until, at the end of a week, the amount is back to normal.

Weaned bunnies should be fed free-choice. All they will eat.

Bucks should receive from 4 to 6 ounces. Adjust to weight. A fat buck becomes a lazy, poor breeder.

Observe your rabbits at feeding time. Pick them up. Base your feeding practices on their physical condition. It is important to be consistent in your feeding routine. Some rabbit keepers feed in the morning and some at night; others fill the hoppers twice a day. Whatever your feeding time, stick with it. We are all creatures of habit, no less so is the rabbit.

Feeding Equipment

You can buy a sturdy, flat-bottomed porcelain crock for $2.50. A metal feeder that fits through a hole cut in the side of the cage and takes up less room and can be filled from the outside costs about $4.

The metal-type feeder is handier and can't be tipped over. Also, the rabbits cannot climb inside and thus soil or spoil the feed. If you are feeding babies, make sure the lip of the feeder is no more than 3 or 4 inches from the floor so that the bunnies can reach it.

Rabbits Must Have Water

Providing a constant supply of fresh, clean water to your rabbits is just as important as feeding. Water helps rabbits digest their dry feed and aids in the balance of body metabolism and vital functions.

There are several options here. A non-tipable porcelain crock costs about $3. This should be cleaned daily. There is a metal waterer on the market that fits through a hole in the cage. The large pan goes inside the cage for drinking ease, and the small cup outside holds the tip of an inverted, quart-sized soda bottle, which is fastened to the side of the cage by a wire clip. This system works on the principle of air pressure and demand and costs about $4.

You can also buy a standard rabbit bottle, a 32-ounce plastic jug that fastens on the outside of the cage with a holding wire and has a drinking tube which goes through the wire of the cage at a convenient height for the animal. When not being drawn upon, the valve at the end of the tube shuts off, automatically, with no drip. This bottle costs $4.

I would not use tin cans for either feeding or watering. If they tip over, they can cause waste or unneeded dampness.

In areas with sub-freezing winter temperatures, the water will freeze and you will have to thaw out the open crocks or make sure the bottles don't crack.

If, after a period of time, you are raising many rabbits and have a dozen cages, you may want to invest in an automatic watering system. This involves tapping into your existing water supply with plastic pipe leading to a reduction tank, which will drop the normal 40 PSI to about 4 PSI. Small pipe lines lead to each cage from the holding tank. Automatic valves with trade names like Drinxall or Dew Drop are installed at the ends of these lines within the cage, and when the rabbits have had their fill, the water supply stops. In northern climates, it would be best to wrap the pipe with electric heating tape.

Breeding Your Rabbits

Breeding may sound complicated, but it isn't. In this most important matter, you are a mere go-between, a courier. First, make sure the doe you have chosen to breed is of good size, well developed, weighs at least 5 pounds and is five or six months old. The buck should be vigorous and be six months old.

All you have to do is pick up the doe and carry her to the buck's cage. *Always take her to him.* Put her inside and wait a minute for nature to take its course. The buck should approach the doe immediately, and, after some preliminaries, mount and service her, and in a matter of seconds it's all over.

Unlike most domestic animals, the female rabbit is polyestrus, that is, she has no regular cycle of estrus. She doesn't come into heat or season on a regular periodic pattern, but, rather, ovulation is brought on by the stimulation of the mating activity itself.

Although does can be successfully bred at least 13 out of any given 16 days, if the doe fights the buck and refuses to accept his advances for several minutes, take her away and try again another day. Do not leave her in with him as they may fight and injure each other.

If, the next day, when you try breeding them again, the doe accepts the buck, and the service is completed, remove her from his cage immediately afterwards. Within five to ten hours after the first mating, some rabbit raisers bring the doe back to the buck for a second breeding to take advantage of the chance to fertilize any ova (eggs) released at the end of the ovulation period brought on by the first mating. This is called double mating.

Now is the time to begin keeping records, which you have to do to raise rabbits successfully. Write the breeding date on an index card and place it on or near the doe's cage.

After the doe has been bred, feed her the normal ration of pellets for the next two weeks.

Is She Pregnant?

If you have a small hanging scale with a pan, you can weigh the doe the day she is bred and make a note of it on her card. Two weeks later, weigh her again. If she has gained a pound or two with just her usual ration of feed, she is probably pregnant.

You can also put her back with the buck again and if she fights him off, she is *probably* pregnant.

The best test of all is called palpation. Pick the doe up by the nape of the neck and place her on a level surface with one hand and slide the other hand under and between her hind legs. With your fingers under her abdominal cavity just ahead of the pelvis, feel very gently for the signs of pregnancy. If she's pregnant, you will feel small, marble-like shapes in her belly. These are the embryos or fetuses of the babies to come.

If you don't feel anything at all, put her back with the buck for re-breeding, you won't waste two weeks waiting for her to kindle and be disappointed when she comes up empty.

If she is pregnant, let her alone in her hutch without disturbance other than feeding and watering for the next two weeks.

Gestation Period

After they are bred, you have at least a month to get used to your new rabbits while they to get used to you and their new surroundings. You will have established a steady routine of daily watering, feeding, cleaning and observing them.

The gestation period of the rabbit ranges from 28 to 35 days, but averages 31 days. Check the individual record card that you are keeping on this particular doe so you can tell when she is due to kindle, and two days before *that date* put a nest box in her hutch.

The Nest Box

This is the box in which the doe will give birth, and it will be the home for the bunnies for the first three weeks of their lives. For medium-sized rabbits, this box should be about 10 inches wide, 8 inches high and 18 inches long. *Remember,* it must fit through the door of the hutch. It can be made of plywood, masonite or wire. It is not recommended that the box have a top as body heat from the litter can cause dampness in a covered box, and lead to serious health problems.

With a nest box of the aforementioned dimensions, it is not necessary to leave one end open for the doe to enter. However, if you build your own box, you may have one end of the box about 6 inches high, instead of 8, for easy access.

Place dry straw or shavings on the bottom of the box for bedding. All-wire nest boxes, which you can buy for about $8, have a cardboard disposable insert for lining the bottom and sides of the nest. The liner costs $1, but makes for excellent sanitation as it is discarded after the bunnies leave home and replaced by a new one before the next litter. The floor of a wooden nest box has to be thoroughly cleaned and disinfected between litters.

There is also a trend toward having the nest box recessed in a space cut in the floor of the cage, so that the top of the nest is flush with the bottom of the cage. This is said to prevent injury to the doe's udders when she climbs in and out. Also, any bold bunnies can fall back easily into the nest if they venture out.

The day before she is due to give birth, the doe will begin to pull fur off of her body and add it to the bedding that you have already provided in the nest box. Keep strangers, small children and pets away from her hutch at this time.

When She Kindles

Leave her alone! The day *after* she has given birth, examine the litter. Count the babies (who are born naked and with their eyes closed), write down the number of live ones on the doe's record card and remove any dead or deformed bunnies and dispose of them.

Although some first-time mothers may be nervous, the maternal instinct is very strong, and she will nurse them 90 percent of the time. If the doe has eight nipples and eleven bunnies, you can either destroy the smallest ones or bring them into your house, put them in a warm box and feed them a formula of milk replacer with an eye dropper. There is no rabbit milk replacer on the market at present, but consult your local vet for advice. Any milk replacer suitable for canines or felines will do the job.

If you have two does that kindle at about the same time, and one litter is larger than the other, take the extra bunnies from the larger group and put them on the doe with the smaller litter. Do not mix babies with an age difference of more than four or five days. (The smallest ones will die of starvation.) Another option would be to wrap the extra bunnies warmly and give them to another rabbit keeper with a doe that has recently kindled a small litter.

After the Doe Has Kindled

Cut back on her feed immediately. Reduce her feed ration to one half of what you have been giving her. In other words, if she has been getting 8 ounces per day, cut back to 4 ounces. This will tend to prevent mastitis, caked udder or milk fever, which is caused when a heavy producer has more milk than the little babies can consume at first. Complications from too much milk can be fatal to the doe.

The Litter

You should observe the new bunnies each day, remove any others that have died and see that they stay together in the nest box. Put any little adventurers back if they have wandered away. Their eyes will open after nine days, and they will be growing fur all the while.

If they are born in very hot weather, make sure they have adequate ventilation; this is when a wire nest box works best. If it is extremely cold, you can hang a heat lamp and train it on the top of the nest. Or put it under the box at a safe distance.

Feeding the Doe After Kindling

The day she gave birth, you cut her feed ration in half. Now that the babies are beginning to drink more milk, I would increase her feed gradually ½ ounce per day, until, at the end of a week, she is back to the normal amount. After that, put her on full feed. That is, keep the hopper full and give her all she can eat. She's now eating for nine. As always, provide plenty of fresh, clean water. The doe's milk production can range from 3½ ounces per day to a high of 7 ounces daily, for about the first three weeks. Then it peaks and begins to decline.

Feeding the New Bunnies

After about three weeks, the babies will start to venture out of the nest and begin to nibble on their mother's feed pellets. The protein content of the doe's feed ranges from 14 to 17 percent, which is adequate for her. But the babies are growing tremendously in bone, muscle and size at this stage and they would benefit if you supplied them with a ration of higher protein content. There are special bunny feeds available with up to 25 percent protein and the pellets are smaller and easier for the little ones to chew. This feed costs a few more dollars per 100 pounds than the regular ration, but can be worth it.

You would then provide a creep feeder, which is made of metal and has openings just right for the bunnies but too small for the doe to reach into. She doesn't need that much protein. I would have them on full feed—no limit. At this time, their gain in size and weight is most efficient as they are getting both milk and solid food. Creep feeders cost about $4.

When all of the young have left the nest box, after three or four weeks of age, you can remove it and give them more breathing space.

Do Not Make Pets of Your Rabbits!

There is nothing in this world as cute and adorable as a fluffy little bunny rabbit, so soft, warm and cuddly when you hold it in your arms. *But,* if you plan to raise rabbits to provide meat for the table, do not become too attached to them. This is easier said than done, but most important. I would not even give names to the mature breeding does, but give them numbers or go by color combinations (hard to do with all-white rabbits). It is almost impossible to butcher a pet rabbit, especially if your kids carried her around the backyard all summer.

You can always rationalize, "I can keep this rabbit for six or seven years as a pet, and she will grow bigger and get older and eat more and perhaps give me a little affection. She won't bark at strangers, guard the homestead or catch mice and rats. Someday she will grow past the stage of providing edible, tender meat. She will eat me out of house and home what with pel-

Rabbits should weigh about 4½ pounds at slaughtering.

lets now costing about $15 per 100 pounds. And then she will die of old age in the end, anyway."

Give them plenty of tender, loving care, but don't make pets of them. If this is difficult for you, remember, for all living creatures, man and beast, there is a time to live and a time to die. You are raising these rabbits for food. You have invested time, money and effort in them. When the young ones have reached fryer size, better than 4 pounds at the age of eight or more weeks, it's their turn to provide for you.

There is a nice lady who lives in my town who started out with two rabbits a couple of years ago. She made pets out of them. She won't let her husband kill them for meat. She only breeds them in late spring or early summer and then rests them all fall and winter. Sometimes, at Easter, she sells a few to neighborhood children. Somehow, the kids' parents always seem to return them a couple of months later. You guessed it. She now has 68 pets in her backyard. And more are due soon. Expensive hobby.

Re-Breeding the Doe

If you are planning to get into heavy production, you *can* re-breed the doe when her litter is four weeks old. However, as you are probably new to the business, I would suggest that you wait until the young are at least six weeks of age. Having nursed eight bunnies for six weeks, she will not be fat, and a slim doe will be less likely to have conception problems. You have to make a decision at this point, when and how are you going to wean the babies?

Weaning Time

To wean means to remove the young from the mother, i.e., the source of milk. Using a pan scale, weigh the babies. If they average from 2½ to 3 pounds at six weeks, you are right on target. If they are 2 pounds or less, I would leave them on the mother, with full feed, and wait until they gain a bit more. As a first-time rabbit raiser, perhaps your minor errors in feeding and management have resulted in the bunnies not making fryer weight in eight weeks; it might take them ten weeks. You can always keep the whole family together until the young are heavy enough for butchering. When the doe's milk supply begins to dry up, she will discourage their attempts to nurse and wean them herself.

There is a feeding problem here. While you want the babies to be on full feed at weaning time, the doe should be cut back to her normal feed ration if and when the bunnies are removed so she won't produce too much milk until she dries up. Separate cages make this much easier and this is when a fourth cage comes in handy.

When you take the babies away from the mother, do it gradually, removing the biggest first, over about a week's time. You must plan your re-breeding and weaning schedules carefully, in order that the doe can be alone in her own cage and have at least two weeks' rest before she kindles again.

Reaping the Harvest

When the young rabbits have reached proper fryer weight, they are ready for slaughter. You could pay a butcher to kill them for you, but this will add to the cost of raising your own meat supply. I don't enjoy killing anything and don't know anyone who does. But it is necessary.

It would be a good idea to ask an experienced rabbit keeper to show you how to kill and dress out a rabbit. You may have to psyche yourself up for the job the first time, but it has been my experience that the poignant sadness disappears quickly. In time, you won't even remember which rabbit you had for what dinner, just how good it tasted. After a while, the succession of past rabbits will be only a fleeting blur in your memory.

Methods of Killing

1. Hold the rabbit by the hind legs with one hand, and with the other, stun it with a quick blow to the head. Cut its throat with a sharp knife and hang it by the hind feet directly over a bucket or plastic bag to catch the blood. Cut off the head.

2. Hold the rabbit by the hind legs with one hand. With the other hand, grasp the head and ears and with a quick movement, jerk the head down and up. Be swift. This is not a loving procedure; you are breaking the animal's neck. Cut its throat and let it bleed into a plastic sack or bucket. Remove the head.

The correct way to hold a rabbit when killing by breaking the animal's neck.

Dressing Out

Spread the hind feet at least a foot apart and tie them to nails in rafters overhead. Just below the hock (hind leg) cut the skin in circular fashion around the leg. Make an incision from that circular cut, on the inside of each hind leg, to the anal opening. Then, starting at the hock where you made your first cut, grasp the skin with your fingers and pull down, peeling it completely off the now headless body.

With the animal's belly still facing you, make a cut from the rear end, circling around the anus and genital organs, all the way down as far as you can go until the knife crosses the abdomen and rests against the rib cage.

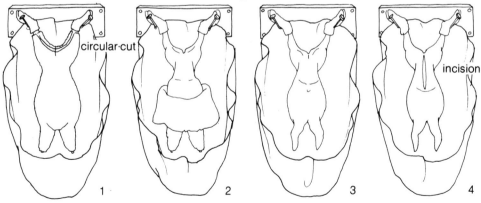

1. Tie the hind feet to nails. Then cut the skin around each leg in a circular fashion. Make an incision from each leg cut to the anal opening. 2. Grasp skin with your fingers and peel it away from the headless body. 3. The carcass with the skin removed. 4. Make a cut in the belly, being careful to cut through only the outer layer of membrane.

Cut thinly through the outer layer of membrane only. A deep cut will penetrate the intestines and inner organs and you don't want to do that. Spread the sides apart, put your hand in the cavity and remove the entrails. Some people save the heart and liver and others discard everything. Your choice.

Remove the hind legs from the body and take off the front legs. You can cut the back into several pieces. Remove any stray hair from the carcass and rinse the pieces in cold water. You will have from 2 to 2½ pounds of delicious, high-protein meat that you can eat immediately or freeze for future use. You don't have to hang it for tenderizing. You can fry it, broil it or barbecue the meat. Or use it in a casserole.

Using the Skin

In my area, a rabbit pelt (fur) is worth about $.50. Commercial processors who purchase live rabbits prefer the white breeds because the pelts can be dyed any color. Years ago, when ladies' winter coats were made of rabbit fur, it was fashionable to call it cony or coney.

You can cure the pelt yourself with a little time and effort. With the edge of a dull knife, scrape all tissue and fat from the inner, fleshy side. Rub generously with plain old table salt over the entire surface. Stretch it out on a flat board and tack it down around the edges of the skin. Keep the pelt in a shaded place. When it's thoroughly dry, the surface will be hard and rough. Using a bar of pumice stone, available at most drug stores, work it over until the skin is soft, smooth and pliable.

Rabbit Manure Makes Excellent Fertilizer

Ever since you brought that first rabbit home and fed it, there have been droppings beneath its cage. In regard to its fertilizing properties, I would rate rabbit droppings just above sheep manure and a bit below chicken excrement. However, fresh chicken manure can burn many plants, but rabbit droppings can be applied, fresh from the donor, anytime and anywhere. You can spread it on flowers, a vegetable garden or the lawn. As a more general all-purpose fertilizer, however, the rabbit manure would be best added to your compost pile, along with leaves, grass clippings and vegetable gleanings.

Turning Worms into Profit

The rabbit manure will be a perfect breeding place for those long, red, active worms that are in demand by fishermen. They will not only cultivate the heap into rich humus, but you can sell them by the dozen to fishermen.

Disease Prevention

It's much easier to prevent disease than to try and cure it. One way is to feed your rabbits right, provide a constant supply of fresh water and keep their cages and housing area clean. Disinfect the nest box after each litter. Clean the cage before the new bunnies come out of the box. You can use a small propane torch to burn hair and other accumulated trash off of the wire of the cage. Poison bait stations for rodents should be maintained at all times if your rabbits are housed indoors. If flies become a problem around the hutches or manure heap, there are many commercial sprays available. When you spray, be careful not to contaminate feed or water dishes.

Sparrows and other common birds can cause disease if allowed inside a shed or garage where rabbits are kept. Dogs, cats and other pets can bring disease in with them and also cause stress. They should be kept away from your rabbitry.

Rabbit Diseases

As a small-scale rabbit keeper, with first-class livestock and a squeaky-clean operation, you should not be troubled with disease. It's usually mass production with too close confinement and unsanitary conditions that causes disease in any animals.

Observe your rabbits closely every day. If they are off their feed or if their feces are not normal—that is, loose instead of hard, round, little balls, these are danger signals. The worst threat to a rabbit keeper is a disease called *mucoid enteritis*. This usually affects young rabbits between one and two months old and the symptoms are a bloated body and violent diarrhea. They can be treated with broad spectrum antibiotics. The cause of this disease is unknown and chances of recovery are slight.

Over the long run, you may have a doe become ill with *mastitis* (caked udder, milk fever). This is usually caused by her having too much milk or injuring herself on a piece of sharp wire. You can prevent this by paying careful attention to her feed rations at kindling time and when the young are weaned. And check the interior of the cage for hazards. The doe can be treated with a shot of penicillin. There is much conflict of opinion among experts as to what to do with the young of an affected doe. Some say you should massage the teats and force her to nurse. Others claim you should put the babies on a foster mother immediately. However, it is very possible that they could infect the foster doe with the bacteria. There is no easy answer.

Rabbits are susceptible to many other diseases such as pneumonia, mange, coccidiosis and ketosis. You can always seek advice from your rabbit-keeping friends or consult a veterinarian who is knowledgeable about rabbits. Whenever a vet visits my farm, the bill usually comes to about $30. If he treats one $10 rabbit, it's a case of diminishing returns. Obviously, if there is a health problem that threatens the whole rabbitry, call in the vet, posthaste.

About the only disease they can transmit to humans is ringworm. This will appear as a loss of fur in scaly patches anywhere on the body. If you decide to doctor the animals yourself, wear gloves and burn them afterwards. You can treat the affected animal with alcohol, medicated sprays or drying powders.

When to Cull Your Rabbits

Domestic rabbits can live for at least six or seven years. Their most efficient and productive years are the first three. If a doe loses her first two litters, or can't feed them, I would cull her. If a buck is a shy breeder or begins to sire small or unhealthy litters, out he goes! Unless you are going to keep him or her for a pet, you have to be callous; don't let your emotions overrule your common sense.

Markets for Extra Rabbits

If, after a period of time, you find that you have more rabbit meat than you can eat, there are several ways to dispose of the extra animals.

1. You can sell dressed meat. At my local supermarket, frozen rabbit, presented in an attractive package and averaging 2½ pounds, sells for $2.29 per pound. You can use store prices as a guide for what you will charge. If you begin to sell the meat in any quantity, you can expect a visit from your county health inspector. There are rigid requirements in regard to slaughter-house facilities, including the use of stainless steel sinks. The department of weights and measures will test your scales for accuracy. Many states require a license for this type of operation.

2. You can sell breeding stock to other rabbit raisers. If, after culling, you have a group of healthy, young does that you don't want to keep, you could start a neophyte off on the right (rabbit's) foot, so to speak.

3. There is a large demand for rabbits from research laboratories. However, you have to supply this market with a quantity of animals of consistent size and quality. Also, you have to be licensed by the U.S. Department of Agriculture. For details, write: Animal & Plant Health Inspection Services, Veterinary Services Program, U.S. Department of Agriculture, Washington, D.C.

References and Sources

Raising Rabbits the Modern Way, Bob Bennett, Garden Way Publishing Co., Charlotte, Vermont, 1976. 158 pages of invaluable information on raising rabbits by an experienced breeder and master showman.

Rabbit Book, Ralston Purina Company, Checkerboard Square, St. Louis, Missouri. 26 pages of general information; free and an informative little booklet.

Rabbit Cage and Housing Plans, Ralston Purina Company. An informative, free brochure with space-efficient cage arrangements utilizing existing or new buildings.

Rabbits, Carnation Company, Milling Division, 1700 Potter Avenue, Kansas City, Missouri. (West Coast address: 5045 Wilshire Boulevard, Los Angeles, California.) 39 pages of excellent information. Free.

Rabbits, published by Countryside Publications, Ltd., 312 Portland Road, Highway 19 East, Waterloo, Wisconsin. A magazine chock full of good information and helpful tips on raising rabbits.

Raising Rabbits, Farmers Bulletin #2131, United States Department of Agriculture, United States Government Printing Office, Washington, D.C. Send for the catalogue which lists titles and quotes prices.

The Journal of Applied Rabbit Research, published quarterly by the Oregon State University Research Center, Corvallis, Oregon. Up-to-date information on the rabbit industry.

Tan Your Hide! Phyllis Hobson, Garden Way Publishing Co., Charlotte, Vermont, 05445. 1977. A fully illustrated, step-by-step guide to home tanning; 144 pages.

CHAPTER 6

PIGS

The first pigs were introduced to what is now the continental United States when the Spanish explorer, Hernando De Soto, landed on the coast of Florida near Tampa Bay in 1539. Along with soldiers and horses, his cargo contained 13 hogs. De Soto journeyed from Florida through Georgia, the Carolinas, Tennessee and Alabama, to the Mississippi River, in search of gold and silver, with the hogs trailing along. He didn't find any treasure, but by the time he arrived at the Mississippi River, three years later, the herd of hogs is said to have numbered 700 head.

This speaks well of the determination of De Soto and his men, as hogs are notoriously difficult to drive, and just the thought of his expedition is mind-boggling. It also speaks well of the prolificacy of pigs. One female hog (sow) can have two litters of from eight to twelve little piglets within a year. Twelve months later, the females from the first litter can bear their young. A year after that, the pigs from the second litter can farrow (give birth) and so on. Everything being relative, then, the offspring of just one, well cared for, modern-type sow could reach well over 700 pigs within three years. I am speaking of pigs being raised under the best of management and conditions. De Soto's hogs underwent the worst of conditions, being herded through swamp and forest in an uncharted land. It's a wonder any of them survived. That they did go forth and multiply was undoubtedly due to their diet, which would have consisted of nuts, berries, ferns, green shoots, grubs, roots, mice, moles, eggs, worms, snakes and anything else that they could catch or poke their snouts into. Hogs, being omnivorous, will try almost anything once.

Rapid Growers

In relation to their body weight and size, pigs grow more rapidly than other farm animals. A piglet weighing only 3 pounds at birth can reach a weight of 200 pounds in five months. If the thought of growing your own hams, pork chops, rashers of bacon and barbecued ribs appeals to you, here is a system for producing a pig of good slaughter weight in about five months. This pig will dress out to about 150 pounds and provide you not only with chops, roasts, bacon and hams, but also liver, heart, kidney, tongue, head, hocks and tail. You can make chitterlings from the intestines, fry the ears for crunchy snacks, make pudding from the blood and pickle the feet. There is a great deal of variety to be found in the eating of a hog.

It should also be said that in order to reach a weight of 200 pounds in five months, the pig must be fed an expensive growing ration such as those prepared and packaged by commercial feed mills. However, if time is not of the essence and you can wait a month or two more, the pig can be fed on kitchen gleanings, fruit and vegetable cullings and root crops. It will take longer for the pig to gain the desired weight, but the meat will taste just as good, and the total cost will be significantly less.

Weanling Pigs

This system is based on your purchasing at least one weanling pig in early summer and feeding it a commercial ration for about five months. If you can supplement the feed with kitchen cast-offs, edible garbage, fresh greens, orchard and root crops, so much the better. As a novice, you will have to pay an experienced person to slaughter, de-bristle and dress out the hog for you the first time. Observe the procedure closely and consider it a hands-on training project. Don't be afraid to get your hands wet or dirty.

Before you buy a pig, make sure that someone in your neighborhood is available to do the slaughtering and dressing job at a reasonable cost (up to $30). This will include the killing, scalding, scraping of bristles, splitting the carcass in half, and cutting it up into recognizable roasts, hams, chops, bacon sides and spare ribs.

The housing need not be elaborate; any shed, barn or three-sided shelter will do. In the summertime, all a pig needs is a roof to keep the hot sun, wind or rain off of its back. It is my belief that you should have a yard or outside pen available for the pig or pigs, rather than keeping them in total confinement. The exposure to sunshine, fresh air, and a chance to root in the soil can make a world of difference in the growth of a healthy pig. One roll of hog fence, 3 feet high and 330 feet long, costing about $100, should suffice to contain one or two pigs and afford them plenty of room to root to their hearts' content. The roll of wire will fence in an area approximately 80 feet square. If you don't have that much room available, you can get by with much less. Even a jerry-built board fence, with dimensions of 10 by 10 feet, or 20 by 20 feet, will expose the pig to that much needed vitamin D. The important thing to keep in mind is: the smaller the pig enclosure, the more likely it will turn into a mud hole.

An eight-week old, 30-pound weaned pig will cost about $35. You can buy 20-pound, weaned piglets, about one month old, for $1 pound. As a beginner, it might be better for you to go with the 30-pound pig as the mortality rate for pigs from birth to 30 pounds averages about 25 percent.

If you have to buy all of the feed and don't give your pig any supplemental goodies except a watermelon rind on the 4th of July, it will cost at least

$70 to raise him or her from 30 pounds to good slaughter weight. If you can supplement the purchased ration with home-grown produce, you can reduce the total feed costs by 10 or 20 percent.

Rather than buy new equipment, I would suggest that you improvise and use any old pans, buckets or vessels that will hold water and a few pounds of pelleted feed. For years, I used old turkey-roasting pans that had seen better days, and they did the job nicely.

If you already have a shed, barn or other outbuilding, your start-up costs will be about $175—for one pig, a roll of hog fence and enough feed to bring the animal up to slaughter weight. However, $100 of this cost is for the fence wire. It is not imperative that you buy this fence, but if you do, it can be used over and over again for many years. If the start-up cost seems excessive, I would suggest that you check the price of pork at your supermarket. If purchased at the store, the comparable quantity of hams, chops, roasts, bacon and ribs that your pig will provide you with could cost at least as much and not compare with the taste of your home-grown product.

You will spend about 15 minutes twice a day, every day, feeding, watering and observing your pig or pigs. Your children can help you with the chores and at the same time develop a sense of responsibility.

Popular Misconceptions

There are a lot of misconceptions about pigs and I'd like to clear up some of them. Pigs are not filthy by choice. They are as clean as you will allow them to be. While we often see pigs standing in dirty hog-wallows in order to keep cool and stave off insects on a hot day, they'd much prefer to immerse themselves in fresh, clean water.

They are the only domestic farm animal that will not foul their own nests (I include horses, cows, sheep and goats) if you provide them with an alternative place to take care of business.

We cook and prepare pork for eating much longer and more carefully than any other meat because of the chance that the trichina (worm) may be present in the raw flesh. It can burrow into the muscle of man in the larval state and cause the disease, trichinosis. There is little doubt that diseases like trichinosis were the basis for the prejudice of Semitic peoples, Arabs and Jews, in their laws forbidding the consumption of pork. And there's little wonder, considering the unsanitary conditions that pigs were raised in way back then. But, in this day and age, with enlightened management and clean surroundings, there is no reason to be suspect of pork, provided that you still cook it long enough.

The theory has also been offered that pigs and their keepers, the swineherd or settled farmer, were looked down upon by nomadic people in the early stages of their history. Jews and Arabs herded goats, sheep and cattle, herbivorous ruminants all, and the animals could exist on whatever sparse grass or browse they found along the way. De Soto's pigs survived the jour-

Young Yorkshire sows.

ney through swamp and forest because of the food they could scavenge. The chances of a pig surviving a desert pilgrimage are poor.

Pigs are said to be greedy and eat like . . . pigs. Well, given a cloven hoof and no thumb, how are they supposed to eat? Being the fastest growing of all farm animals, in relation to birth weight and ultimate size, they have to have a good appetite in order to gain a couple of hundred pounds in about five months, and that's the desirable goal when you are raising pigs for the table.

In comparison to pigs, a newborn kid (goat) will weigh perhaps 7 pounds and may reach 50 pounds in five months. The single lamb can weigh from 7 to 14 pounds at birth and from 80 to 100 pounds at five months of age. The calf (of a beef breed) will weigh from 70 to 85 pounds at birth and reach a weight of 350 pounds in five months.

Some breeds have short, stubby snouts, and some have very long protuberances. Whatever the length, the pig's snout is a miraculously engineered instrument for rooting, up-rooting, digging, shovelling and mucking about for food. They literally slurp and suck food into their mouths. The clever, disc-like shape at the end of the snout is made of a sturdy, flexible tissue attached to strong muscle fibers that can rotate it up and down and sideways, at almost any angle, to find tasty morsels like truffles and toadstools.

Check the Zoning Ordinances in Your Town

While you don't need much acreage, you will probably have to live on the outskirts of a village or in a semi-rural area in order to keep pigs.

Just because a farmer down the road raises pigs, it doesn't necessarily follow that you will be allowed the same privilege. There may be a "grandfather" clause in your town code to the effect that agricultural occupations in existence before certain zoning laws were enacted can be carried on.

Pigs don't really smell, especially when kept outdoors; it's their pens or wallows that can have a strong odor. I have to smile when city visitors turn up their noses at pigs. The pig is a useful, productive animal and usually confined to a specific area. You can always walk away from their pen. Better than dodging dog feces in a big city, pooper-scooper laws notwithstanding.

It might also be a good idea to let your nearest neighbors know what you are planning on raising in that little pen out back. Perhaps you can promise them a rasher of home-grown bacon, cured and smoked, or invite them to a barbecue of country ribs in the fall after you've successfully raised your first pig to slaughter weight. They might even be your best customers and buy a half or whole pig when the time comes.

When to Buy Pigs

Late spring or early summer is the most opportune time to buy piglets. June would be ideal as the pig would be ready for butchering by the time the crisp weather of fall sets in. Some commercial hog breeders aim for three litters per sow, per year, but many more pigs are born in the springtime. In this season, you would have a much better choice. Also, temperate weather would make it easier to raise the pig or pigs. From the day you bring them home, they could be kept outside, given some sort of shelter.

Where to Buy Your Pigs

If possible, purchase them from a local farmer. There is nothing basically wrong with buying at a public stockyard auction, but the chances of the animals being exposed to disease at an auction are far greater, and young pigs are susceptible to many diseases. If you buy from a local farmer you can check out his operation, take a look at the mother pig and satisfy yourself as to the care the pigs have received.

I would also suggest asking a friend to accompany you, a person knowledgeable about pigs. Most of us know one of those legendary farmers down the road, one of that diminishing breed, a sturdy part of the 4 percent of the people who feed our nation and to whom time isn't always money. He still has time, on a summer's morning, to watch a red-shouldered hawk drop with folded wings on a hapless field mouse, count the wild Mallard ducklings in the pond and help a greenhorn avoid a pig in a poke.

Buying Your Pig

Among the most popular breeds of swine in North America are the Chester White, Duroc, Hampshire, Landrace, Poland China and Yorkshire. The Chester White, Landrace and Yorkshire are white. The ears of the Yorkshire stand up erectly. The Chester White has small lop-ears, and the Landrace has long, floppy ears that almost cover his eyes. Don't worry, he can still see you coming from a long way off and always knows when it's chow time. The Hampshire is black with a kind of white collar that extends over his shoulders and front legs. The Duroc is of reddish color, and the Poland China is black with white spots. There is also a Hereford pig with the same white face and red body as his bovine counterparts, those Hereford cows that the rustlers are always rustling in cowboy movies. And there are dozens of hybrid pigs, with names like Minnesota No. 1 or Beltsville No. 1. All of them are good prospects. However, your choice will be limited to what's available in your area. As with all domestic farm animals, there is no *one* best breed, there are only superior strains and families within a breed.

If you buy from a local farmer, put your order in early and ask him to let you know when the pigs are weaned (removed from the mother) and ready for sale. This way you can have first choice and not be forced to settle for the remainder of a picked-over litter.

Look for the biggest and most active pigs in the group. Don't buy the runt of the litter, and avoid the droopy-looking one standing off by itself in a corner or any that have swollen bulges underneath their bellies indicating the possibility of hernias.

I would suggest that you buy at least two. Pigs love company and there is a competition factor involved. In competing for food, they will eat better and gain faster, and that's exactly what you want. If you sell all or part of the second pig when he is of slaughter weight, it will help defray part of your expenses, and feeding two is no more time consuming than feeding one.

Whether they are males or females doesn't matter. The gilt (female) will gain just about as fast as the male, and the meat tastes exactly the same. If you buy a male, it should be a barrow (castrated). Their wolf or needle teeth will have been clipped. And they should have been given a shot of iron, shortly after birth, to prevent anemia. The pigs should be about eight weeks old and weigh around 30 pounds. A 30-pound pig has a much better chance of surviving than a smaller, younger weanling, as the most vital period of time in a pig's life is the first few weeks. *A very important note: the pig must be accustomed to eating solid food.*

There is another route that you can go and that is to buy a mature or aged boar (male pig). He will be from four to six years old and weigh at least 500 pounds. In this venture you will get the most meat for the least price. The

Four-week-old Yorkshires.

catch here is that the boar cannot be slaughtered immediately as he is an entire male. He has been used for breeding sows, and the flesh would have a strong taint to it. He would have to be castrated and then kept for several months until his system had been purged of any lingering traces of his erstwhile masculinity.

A vet could charge up to $30 for the operation. An advantage here is that you won't have to feed him heavily in order to make him grow. He's already grown to full size. You would feed him a maintenance ration, high in fiber and bulk, low in protein. As an aged boar at a public auction, he would bring bottom dollar and you might purchase him at a bargain price. Or you could buy him from a hog breeder who can't use him anymore. He could dress out to some 350 pounds of edible meat, and the hams, chops and spare ribs would be tremendous in size. I have never gone this route but have tasted the end product and it was very good. Tending the boar won't be as much fun as raising your own piglets, and you won't have the pleasure of watching them grow up. I would suggest that you buy one or two weaned pigs and raise them from scratch.

Housing, Pens and Fencing

The pig can be raised in total confinement, and you can use any type of barn, shed or other outbuilding to keep him in. However, as this system of raising pigs is based on beginning your venture in late spring or early summer, temperate weather affords you the chance to keep them outdoors if you have the space. All they need is a simple structure to keep the wind, rain and hot sun off of their backs. My pigs lived outdoors from the time they were six weeks old until they reached slaughter weight.

I made a very simple three-sided, thatched-roof shelter out of old boards. The shelter was 6 feet wide, 4 feet long and 5 feet high. I nailed boards up the three sides, to a height of 4 feet, leaving a 1-foot gap just under the roof, to provide ventilation on hot summer days. I placed a section of chicken wire across the top, put down a layer of straw about 6 inches thick on top of that and then fastened another strand of chicken wire over the straw to keep it from blowing away. The opening faced the south-east and I used old straw and hay (not spoiled or mouldy) as bedding for them to burrow in.

A simple, three-sided shelter may be fabricated with boards, straw and chicken wire.

The uprights were not driven into the ground and the structure was easy to lift or turn or even move to a new spot if the bedded area became damp from rain. The pigs did not soil their own nest but always traveled some distance away to take care of business. This shelter housed three pigs from the time they were weanlings until they were well over 200 pounds each.

Small "A" frames are easy to build. If they do not have floors, they should be moved to dry ground when necessary, i.e. in damp climates or during rainy spells.

Enclosures. If you have a choice, I would suggest that you keep your pig or pigs outdoors. I believe they are more healthy and gain faster when raised outside.

If you must keep them inside, allow about 50 square feet for each pig. This will give them room to keep their bedded, sleeping area separate from their manuring place. The bedding can be of wood shavings, straw or old hay (not musty or mouldy), and, if the pen has a slope to it, the litter should be placed at the highest level. The floor can be of concrete or wood, the former being the easiest to keep clean. If your indoor pen is very small, it should have windows for ventilation and a bit of fresh air on hot days. If you use a large barn, you can partition off a small area that the pig can use for sleeping quarters.

If there is any space adjacent to the barn or shed, take advantage of it and fence it in. A yard 20 by 20 feet or even 10 by 10 feet is better than no yard at all. It will give them a chance to soak up some vitamin D when they poke their heads out of their house on a sunny day. If the space is small, you can make a solid board fence using scrap lumber. Make sure the bottom boards touch the ground, and, at lower levels, the less gap between the boards, the better.

If you do have space available, one 20-rod roll of hog fence (330 feet) will provide a pen 82.5 feet square, more than enough for just a pig or two. Whenever you fence, keep this in mind: the squarer the area, the less wire needed. A long, narrow field requires much more fencing to enclose than a square field, even though both fields have exactly the same area.

Hog fencing does not have to be high, from 3 to 3½ feet will do. Being rooting animals, pigs have a tendency to go under, rather than over, fences. However, the fence must be sturdy, especially at the bottom. Set posts firmly, at least 18 inches deep, depending on the average depth of penetration of frost in your climatic zone. Posts set too shallow may heave up out of the ground during spring thaws. This will not affect you during your first summer of pig keeping. But, if you are going to do the job, you may as well do it right.

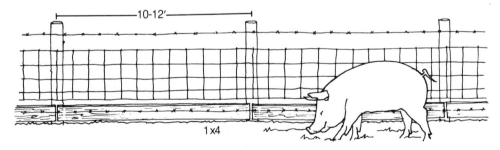

Hog wire, used in conjunction with 1 x 4s and barbed wire, may be used to contain pigs effectively.

If you use hog or page wire, space the posts from 10 to 12 feet apart. Staple the wire at ground level. Always affix fencing on the inside of the posts in a pen or pasture. This way, the posts serve to brace the wire when animals lean on it. The 1-x-4-inch boards (or bigger) should be nailed at ground level, on the inside of the posts. It could be of value to staple a strand of barbed wire, 2 or 3 inches above the ground, on the inside of the posts, to discourage venturesome pigs from escaping.

Electric fencing is often used successfully to contain pigs. Space the posts from 6 to 8 feet apart; the wire requires plastic insulators to isolate it from the posts. Two strands of wire are recommended. Place the first wire about 6 inches above ground level and the second strand about 12 inches higher. Keep grass and weeds trimmed from under the wire so that they don't touch it and ground it out. Electric fencing requires constant surveillance. A 6-volt storage battery delivers a mild jolt, sufficient to keep the pigs within the pen. Once they get the initial mild shock, animals avoid the wire like the plague. One 6-volt battery, costing about $15, will usually last for at least a season or two.

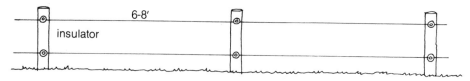

Two strands are recommended, if you use electric-wire fencing.

Once I worked from dawn to dusk building a small pen for three little pigs I was raising. I put the pigs in the new pen just before dark and returned to the main barn to finish other chores. Ten minutes later, I looked up and there they were; they had gotten out and the three of them were staring at me through an open barn door. If pigs can smile. . . . I do think they have a sense of humor, especially when they're young. If they do get out, don't panic and don't chase them. Even if you did catch the pig, there's no handle to grab onto. Rattle a bucket of feed under their noses and they should follow you back home, curling their little tails behind them.

Necessary Equipment

You will need a trough for feed and another for water. Generally speaking, 1 foot of space is required for each pig. The smallest metal hog trough available is 2 feet long and a new one costs about $14. If you are handy with a cutting torch, you can cut an old hot-water heating tank in half and it will serve the purpose. V-shaped or rectangular wooden troughs are easy to build. Soak them in water for a few days, to prevent leakage. Or, you can use old turkey-roasting pans or trays. I find that pigs do better when I supply one trough for each pig, rather than feed them all out of a big trough. There is a bully in every crowd and, if you have more than one pig, the boss pig will hog the trough, while the less aggressive get the short end of the stick.

It's important to have an idea of the weight of your pig during various stages of its growth. A spring-type scale that will provide accurate weights up to 60 pounds costs about $24. You can tie the pig's hind feet together and hang him on the scale hook. Or, cut a section of inner tube and fasten it around the pig's belly and hang it with twine. A hog weight tape is made of non-stretch fabric and is drawn snugly around the girth of the pig, just behind the front legs and completely encircling the body. The reading in inches is converted to pounds and the tape gives a fairly accurate estimate of the pig's weight, plus or minus about a 5 percent margin for error. The hog tape will cost about $2. If you can find a dressmaker's tape around the house, wrap it around the pig's girth. A reading of 39 or 40 inches will show that your pig is of good slaughter weight. The best time to tape your pig is when he is busy eating and doesn't mind if you wrap it around his body.

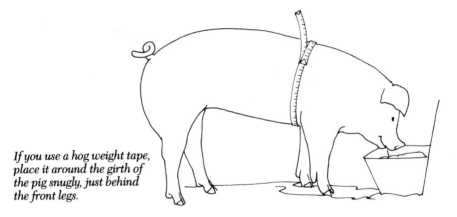

If you use a hog weight tape, place it around the girth of the pig snugly, just behind the front legs.

Feeding

Before you bring your pigs home, ask the farmer what kind of feed they have been eating. If possible, continue using exactly the same ration. It is not good practice to make an abrupt change in the feed of any animal. A change of feed, coupled with the stress of transporting the animal from one. place to another, can cause digestive problems and a possible setback in the pig's growth. In all probability, the piglet has been getting a completely balanced starter ration of 18 to 20 percent protein.

Start out by feeding this high-protein starting ration at the rate of from 1½ to 2 pounds a day. If the pig cleans it all up and seems to want more, gradually increase the amount of feed. If he doesn't finish it and wanders off to chase a butterfly, decrease the amount. At his age and weight, the pig should gain about 1 pound a day. If he weighed 30 pounds when you bought him, he should reach 50 pounds in less than a month. Try to estimate his weight even if you have to hold him in your arms while stepping on a bathroom scale.

When he weighs 50 pounds, you can switch to a growing ration of about 14 percent protein. You will have some starter ration left over and you can mix the remainder of the starter with the grower in order to provide a smooth transition in the change of feeds. By the time the pig has eaten 550 pounds of grower, he should weigh about 200 pounds and be ready for slaughter. In my area, 18 percent protein pig-starter costs $13 per 100 pounds. The growing ration of 14 percent protein costs $10.40 per 100 pounds. At these prices, the feed costs would be about $70.

Some commercial feeds are meant to be fed dry and others to be mixed with water. Even if the one you use is of the wet type, always keep plenty of fresh water available for your pig.

Supplemental Feeding

As a beginner, you cannot duplicate the complete, well-balanced ration that the feed mill puts together in scientific proportion. This ration contains all of the ingredients necessary for the young pig's growth and health: protein, fat, fiber, vitamins and trace minerals. However, you can certainly supplement his diet with vegetables and fruits culled from your own garden or orchard. The wonderful thing about pigs is that they will eat almost anything, and with great gusto. They will eat fish, meat scraps, skim milk, old bread and bananas.

If you live near a dairy and supply your own containers, perhaps you can obtain skim milk or whey. Maybe you can make a deal with the manager of your local supermarket for the bananas, apples, tomatoes, lettuce, cabbage, potatoes and turnips that would otherwise be taken to the town dump. A bakery could provide old bread, rolls and pastry. You could supplement the pig's feed with any of these items and cut your feed bill by 10 to 20 percent. However, this supplemental type of feed should not constitute more than 15 to 20 percent of the pig's daily ration. When pigs are very young, they need the highly concentrated starter ration. Fruits and vegetables will not provide enough protein, vitamins or minerals. Also, their intestines are not ready to handle large quantities of fiber.

If fed heavily, dairy by-products can cause pot-belly, especially when the pigs weigh over 150 pounds. An overabundance of bakery goods can cause constipation in pigs. In the beginning, it would be wise to feed the supplements in moderation.

If pigs have access to a small lot or pasture, they can obtain many of the nutrients they need from the grass and earth. If your pigs are confined indoors, give them a large shovelful of sod, grass and all, a couple of times a week.

Sanitation

If the pig is raised in total confinement, I would rinse his feed and water trough and clean the pen daily. Place the bedding at the highest point in the pen so that urine and fecal matter will drain away from it. Pigs will usually do their thing in the same place every day. Pigs on pasture will spread manure over the field.

Health

In young pigs, scours is probably the most destructive disease. Symptoms of scours are diarrhea, loss of appetite and a dehydrated condition. An abrupt change of feed or stress can cause scours. The contamination of feed by mice and rats can bring it on. Moist or liquid feeds that have spoiled can cause scours. For pigs kept indoors, the best treatment is to follow a strict program of sanitation and provide good feed and plenty of water. Always keep pre-mixed feed in tightly closed containers to prevent loss and contamination by mice and rats.

I would not hesitate to feed older pigs wilted or overripe fruits and vegetables, but shy away from anything that is mouldy, spoiled or rotten, particularly dairy products. Never feed pigs the leaves of the rhubarb plant; these can be deadly.

If your pigs are housed in a building where no other animals have been kept, or penned outdoors on uncontaminated ground, there should be no problem of disease. Most animals have worms, however, and some pig growers recommend worming them periodically. Worms can cause a pig to lose weight and appear unthrifty. Worming medications are available at feed stores and are added either to feed or water. If you decide to worm your pigs, follow the directions carefully, as the amount of medicine to use is generally based on the size and weight of the animal.

Daily Care and Observation

If you only have a pig or two, you will probably hand feed them twice a day. This gives you an opportunity to observe them closely. Does the pig have a good appetite? Does he run eagerly to the trough? Or is he coughing, dull-eyed or listless. Don't get excited if he takes a nap immediately after eating a big meal. This is normal. But if there appears to be a problem, consult a farmer knowledgeable about pigs or a swine specialist associated with your county or state agricultural department. A visit by a vet would probably cost more than one small pig is worth. However, a phone call to a friendly vet might help provide direction in treating the problem.

You will begin by feeding the pig 1½ to 2 pounds of feed a day. If he cleans up all the feed, increase it gradually. Even if you are feeding a wet ration, always provide plenty of fresh water.

Try to estimate the weight of the pig with the aid of a scale or tape. Assuming you bought the pig at 30 pounds, within three or four weeks he should weigh 50 pounds. Mix the leftover starter feed in with the grower feed and increase the ration.

Everything being relative, he should weigh about 100 pounds within another month's time. He will be eating over 3 pounds of feed per day at this point and gaining close to 2 pounds each day.

By the time he approaches 200 pounds, he will be eating over 4 pounds a day and gaining more than 2 pounds daily. These are not hard and fast figures, but offered only as a general guide. As a beginner, you cannot expect to obtain results as good as those of an experienced pig farmer.

Your pig will have eaten about 650 pounds of store-bought feed, plus whatever fruits, vegetables and other goodies you supplied him with. When you estimate that the pig weighs about 200 pounds, it's time to convert him to edible pork. At higher weights, pigs have a tendency to lay on fat and it's the muscle that we eat, not the fat.

Slaughtering

Notify the man who will do the slaughtering that your pig is ready. He will probably tell you to give the animal plenty of water, but no feed, the night before it is to be killed. This will make the dressing-out process less messy.

The animal will be rendered insensible by a stunning blow to the head or else shot between the eyes with a rifle. Immediately afterward, the arteries in the neck will be cut with a sticking knife and the pig will be bled out completely. After that, it will be immersed in a tank of scalding water, of about 145° F., for four or five minutes. The carcass will be placed on a table and the bristles removed with a scraper. It will then be washed and the head cut off.

The body will be cut down the middle from throat to crotch and the entrails and organs removed. The carcass is then split in half and hung in a cooler for a day. The chilling process not only removes body heat but makes the carcass firm and easier to cut up. I would suggest that you observe the whole process closely, and one day you may be able to do it yourself. In any case, the butcher will welcome any help you can offer, whether it's to lend some muscle on the ropes when the pig is raised over the scalding tank, or to help scrape the bristles off of the skin.

Dressing Percentage

A 200-pound pig will dress out to about 150 pounds of edible meat, including the liver, kidneys, heart and tongue. When the carcass is cut up, the ratio of choice meat between the front and hind quarters is about 50:50. The four primal cuts of the hog are the hams, shoulders, loin and bacon belly. A dressed carcass of 150 pounds will provide you with about 50 pounds of ham and bacon. In my area, it costs from $.10 to $.15 per pound to have the hams and bacon smoked professionally.

Besides the hams, bacon, spare ribs, chops, roasts, liver, heart and tongue, the pig offers an almost endless variety of special cuts and treats. If so inclined, you can collect and save the blood for pudding. A gelatinous head-cheese can be made from the pig's head. The animal's intestines can be used to encase delicious sausage made from belly trimmings. Some people roast the hocks and tail, fry the ears and pickle the feet. And there's always the fat for making lard and soap. The possibilities are fantastic.

Manure

Although the odor can be almost overpowering when you get a whiff of freshly spread pig manure, its fertilizing value is somewhat less than that of poultry, rabbits or sheep. If your pig was pastured outdoors, the manure will make next year's grass green. If he was penned inside, the manure, plus litter, is an excellent additive to your compost pile.

References and Sources

Animal Science, M. E. Ensminger, The Interstate Publishing Co., Danville, Illinois, 1969. This text of 1,200 pages covers everything from chickens to horses; there is an excellent section on swine production.

Small Scale Pig Raising, Dirk Van Loon, Garden Way Publishing Co., Charlotte, Vermont, 1978. This many-faceted book of 272 pages is aimed at the beginner, but even the most experienced farmer can learn new things from it about raising pigs. It covers everything from how to buy a pig to curing the meat of the finished product.

CHAPTER 7

LAMBS

Recently, while I was checking out the meat counter at a supermarket, the lady standing next to me whispered, "Now that the holidays are over and all the company is gone, I can have what I really like, lamb." And she put a package of choice lamb chops in her shopping cart. She needn't have been apologetic or have spoken in hushed tones. She should have shouted the praises of lamb out loud! We could live off the fat of the lamb! Well, not really; it wouldn't be wise to live off the fat of anything. However, the fat of the lamb is concentrated largely on the outside of the carcass and easily trimmed off, not marbled throughout the meat like in the prime beef steaks we North Americans are so fond of.

According to nutritionists, the meat and fat of the lamb and adult sheep (mutton) is more easily digested than the meat and fat of either beef or pork. This is one reason why it is highly recommended for hospital patients and the elderly.

This particular lady was middle-aged, and I assumed that she had grown up in a family that had lamb for dinner with some regularity. And yet, I would bet that her own children didn't eat lamb. I would further wager that if a school teacher asked the class how many of them ever ate leg of lamb or lamb chops at home, no more than two hands would be raised. Annual per capita consumption of lamb in the United States is about 2 pounds.

Less than 10 years ago, our per capita consumption was about 4 pounds a year. Somewhere along the way, the custom of eating lamb or mutton fell out of favor. It has been blamed, in part, on the fact that during the first and second World Wars, servicemen were fed mutton imported from New Zealand and Australia. The mutton came from older sheep that had been raised primarily for wool, not tasty meat, and the servicemen didn't like it. There is a world of difference between mutton and young, tender lamb, but prejudices are very hard to overcome.

Another reason for the dwindling interest in lamb is that its price is not competitive with beef or pork. In my area, loin lamb chops retail for $4.50 per pound. Loin pork chops are sold for half of that amount. Shoulder lamb chops cost as much as sirloin steak. No contest there. Perhaps if some fast-food company opened a chain of drive-ins with names like "Lamburger

King " or "Bo-Peep's Kitchen," our consumption of lamb would increase. As it is now, the price goes higher and less is consumed.

In the early 1800s, New York and New England had more sheep than any other part of the nation. Every farm had at least a few sheep. Sheep used to graze on Boston Common. Admittedly, they were kept for their wool and not their meat. People did eat mutton, which is not as tender as lamb and has a stronger taste, but it can be prepared in many delicious ways. By the mid-19th century, the majority of the sheep population was centered in the mid-west. Then, when that area became heavily populated and the value of farm land rose, the sheep industry moved farther west, where there was extensive open range for grazing purposes. It was about this time that the conflict began between cattlemen and sheepmen. The range wars of the west were not fought over the fact that sheep are destructive to the land, although, due to the shape of their teeth, they can crop grass closer to the ground than cattle can; or the fact that cattle won't graze where sheep have been. As a matter of fact, cattle and sheep can do well together on the same pasture. Sheep will eat weeds and browse that cattle won't touch. The range wars were fought because of greed; both the cattleman and sheepman wanted to make money from the open range. Today, the great majority of sheep and lambs are raised in the west, and most of their meat is consumed in the east.

A Workable System

If you have tried lamb, and the memory of sweet and tender lamb chops, leg of lamb or roast rack of lamb makes you hungry just thinking about it, here is a system for raising your own lambs. It is not a quick method, like turning out a 4-pound chicken broiler in eight weeks or a 200-pound pig in five months.

This method requires that you buy at least two older bred ewes (female sheep) in the fall of the year, provide them with some sort of shelter and feed them for about five months, until they have their lambs in the spring. It would be normal if the two ewes produced three lambs, as twinning is common with sheep, and triplets are not at all rare. The lambs would nurse from their mothers until the grass was up in the spring, and, with the combination of milk and decent pasture, within five months after their birth, the lambs would weigh about 100 pounds and be ready for slaughter. It would be about ten months, then, from the time you first purchased the ewes until their offspring could grace your table. In my opinion, this is the only economically feasible method in which to raise lambs, and I'll explain why.

With the exception of about three breeds, sheep are seasonal breeders, and that means that the ewes come in heat in the fall and are bred. The gestation period of the ewe is about five months. Thus, most lambs are born in the spring. Trying to buy one lamb from a sheep farmer in the spring would not make sense. In the first place, it would be nursing and he would

A young lamb foretells the coming of spring.

not separate it from the mother. In the second place, since he has already gone to all the time and expense to feed the mother all winter, he would realize more profit if he raised the lamb himself, turning the ewe and lamb out on his own pasture and letting the lamb have both milk and grass.

If you were able to buy an orphan lamb from a local sheep farmer, you would have to feed it with lamb milk replacer, which costs $28 per 50-pound bag in my area. It would also take about $10 worth of grain to carry the lamb until green-up time in late spring when you could turn it out on pasture. You would have to bottle feed the lamb from four to six times a day for the first three weeks of its life, and perhaps three times a day for the next two months. At that point, your lamb could weigh about 50 pounds and you would have spent $38 for milk replacer and grain. At current market prices of $.60 per pound, you could buy a feeder lamb weighing 60 pounds for $36 in late summer and save yourself all that time and trouble.

If you think that you want to try your hand at raising lambs, keep in mind the fact that it may be several years before you begin to break even on your investment. It is also very important to consider the following factors before getting into the sheep business:

On the plus side, sheep produce two crops: lambs and wool. The most inexpensive way to feed sheep is through good pasture. As a rule of thumb, seven sheep can live on the same amount of acreage that it would take to support one cow.

On the other hand, there are certain conditions that would be unfavorable for raising sheep. A prevalence of predators is one. It has been estimated that well over 1 million sheep are killed by coyotes and stray dogs each year.* Sheep are subject to many parasites, both internal and external. They are less hardy and resistant to disease than pigs or cattle. It is very expensive to provide adequate fencing for them. I would not suggest that you keep sheep with the idea of having them clear unimproved land. If there is a choice between tender, lush grass and browse or weeds, the sheep will go for the grass every time.

If you decide to raise sheep and lambs, and follow this method, you can put fresh lamb in your freezer within 10 months. A 100-pound lamb will dress out at about 45 percent of live weight. Two lambs will give you about 90 pounds of meat—heart, liver, kidneys and bone included. If one of the ewes has twins, that's an added bonus.

You will spend about 15 minutes twice a day, feeding, watering and tending the sheep. You can use an old barn or shed for their shelter. All they need is protection from a cold rain and the wind. You can use a washtub for their water and make a rectangular feed trough out of old boards.

Two, older bred ewes will cost about $75 each. Between them, over the winter, they will eat about 150 pounds of grain which will cost about $30 if purchased at a feed mill. They will eat about 10 bales of hay each. If you can buy hay for $1.50 per bale, that's another $30. A 50-pound block of salt will cost $4. If your pasture is just average, the sheep and lambs will require two acres, and it will take four rolls of woven wire fence to enclose that area. The wire will cost about $100 per roll, for a total of $280. If you space them 12 feet apart, 100 posts will be needed to hold the wire for a fairly square two acres. In my area, softwood, untreated posts with points, 6 feet long, cost about $1 each. Black locust or cedar posts are much better and cost more. Our local supply store sells steel posts for about $3. A box of staples costs $5. Adding it all up, your start-up costs will be at least $600. As you can see, the main cost is for fencing materials. However, the fence wire can last 10 years.

Buying Your Sheep

October or November would be a good time to buy the ewes. If possible, take a farmer knowledgeable about sheep with you. Buy from a local farmer, rather than at an auction barn. Less chance of disease.

Some of the more popular breeds of sheep in the United States are Suffolk, Hampshire, Corriedale, Dorset and Finnsheep. For our purpose, you would buy good grade (not purebred) ewes, three or four years old. Some experts recommend buying yearling or two-year-old ewes, and I won't

---

* See "Lamb Predators," Ray and Lorna Coppinger, *Blair and Ketchum's Country Journal*, April 1980.

argue with that. However, as you are a beginner, it might be better to go with ewes that have already lambed in past years and know the ropes. Some first-lamb ewes won't let their lambs nurse and some ewes disown their lambs. In the beginning, you'll need all the help you can get. And younger ewes would cost more.

When you look at the sheep, make sure they move well and don't limp. A ewe with bad legs or feet will have trouble getting around the pasture and grazing. And when she gets to the grass, she has to be able to eat it. You have to check her teeth. By the time a sheep is four years old, all the baby teeth have been replaced by permanent teeth. Don't buy a ewe with a broken mouth (teeth missing) or a gummer (no teeth at all). To catch a sheep, approach it from the side with its head to your left. Enclose its head and neck from underneath with your left arm and grasp its right hind flank or leg with your right hand. Try not to grab at the wool as you can hurt the sheep or disturb the natural pattern of its fleece. When you've got hold of the sheep, dig your right knee into its left flank, lean backward slightly, and the animal should slide easily to the ground. Then, with your right arm holding the sheep immobile, your left hand is free to open its mouth, pull down the lower lip, and examine the teeth.

Also, check the udder to make sure it has two nipples. Ask the seller when the ewes were bred so that you can estimate when the lambs are due.

Housing

You can use any type of outbuilding, shed or barn to keep them in. All they need is shelter from cold rains and strong winds. The ewes don't need to be kept warm. They are wearing a natural fur coat. If you use an old barn, you can partition it off if you want to confine them to a smaller area. It would be best if there was a pen or yard adjacent to their housing so that they could go in and out as they choose. My own sheep spent most of the wintertime out in the snow, going inside only to get away from the wind. Any gates or doors in their shelter must be wide enough so that they can pass through, freely, without injuring the unborn lamb.

It would be a good idea to use old straw (not musty or mouldy) for bedding, especially if the floor of their shelter is concrete. Before the lambs are due, you will have to provide a small lambing-pen for the lamb and mother. This pen has to be draft free and dry. An area 4 feet by 5 feet will do just fine. The ewe and lamb are kept in this pen for the first few days after the lamb is born. You have all winter to build it. Plywood and 2 x 4 framing are good materials.

Fencing

When you fence for sheep, you are not only keeping them in, but also keeping predators out. The fence should be of 36-inch woven wire with two strands of barbed wire spaced above and one strand of barbed wire at ground level, on the outside of the fence posts. The barbed wire is to prevent predators like coyotes or dogs from jumping over and crawling under the fence.

The more squarely that you can fence a pasture, the less wire is necessary. A long, narrow field takes more wire than a square one of exactly the same square footage. If there is a stream on your land, try to include part of it within the pasture. This will save hauling water by hand in the summer time. Don't forget to make a gate.

Typically, sheep fencing includes woven wire and barbed wire at the top and bottom to keep out predators.

I would not recommend using barbed wire for fencing sheep pastures. You would need about 8 strands, closely spaced together, and the sheep could injure themselves or tear their wool when pushing against it. Electric fence would be the cheapest, but the sheep's wool could insulate the animal from receiving the mild shock and render the fence ineffectual.

Necessary Equipment

You will need a trough for feed and a receptacle for water, and a rack for hay. It is better to keep hay off of the ground when feeding, as there is less waste. See Appendix D for plans for building a trough and simple hay rack. A hoof trimming knife would come in handy. During the course of a year, you would probably trim the ewe's hooves at least twice. This knife will cost about $5. An old washtub can be used for their water trough. Or you can buy a heavy-duty, plastic, 6-gallon water tub for about $7.

About Predators

Over 1 million sheep and lambs are killed each year by dogs and coyotes. Even your own good-natured family pet can turn into a killer. Most dogs will chase anything that runs away from them, and sheep will run because they are defenseless. The dog doesn't have to tear the sheep's throat; just the stress from being chased can cause an older ewe to have a heart attack. You are the shepherd and it is up to you to protect your flock from dogs. Sheep ranchers in the far west use traps, poison baits and sheep *guard* dogs for protection against coyotes. Don't feel smug just because you live on the east coast. The coyote is alive and well in New England.

Feeding

For the first half of their gestation period, all the ewes need is hay, the greener and more leafy the better. Four to five pounds of good hay per day should be sufficient for a medium-sized ewe. Hay is usually fed for about six months, but that depends upon the climatic zone in which you live. They will need hay until the grass appears in spring. Always give them a good feed of hay before you turn them out on pasture. An abrupt change in feed can cause severe stomach upset in any animal. Give them plenty of fresh water and keep salt available at all times.

For the last half of their gestation period, begin to give them grain, about ½ pound per ewe daily. This can be increased, gradually, to ¾ of a pound daily.

For example, if the ewes were bred to lamb in March, and you had purchased them in November, you would feed only hay from November till mid-January. From then until the lamb was born, you would feed a daily ration of grain, along with hay.

Pre-Lambing Observation

Shortly before the ewe gives birth, she will appear nervous. She will lie down, stand up and then lie down again. Sometimes she will turn her face towards her rear end. This is called, "calling for the lamb." It is best to give the ewe plenty of space to move around when she lambs. After the lamb is born, move both mother and lamb to the small pen.

Lambing Time

Toward the end of the gestation period, check the ewes several times a day and night, if possible. Shepherds with large flocks check their ewes every three or four hours, around the clock, at lambing time. Don't upset the ewe, but try to observe discreetly.

If the lamb is born when you are not present, with any kind of luck, the next time you make your rounds, it will be on its feet and nursing, and the ewe will be drying off its coat and sniffing it. The mother will have done all the work.

If the lamb is born while you are present, there are things you can do to help, provided you don't interfere with the mother. As soon as the lamb is dropped, make sure that all mucous is wiped away and cleared from the mouth and nose, so it can breathe. If the ewe doesn't do this, you can use a clean towel or cloth and do the job yourself. If the mother is aloof and not doing anything, rub the lamb's body briskly to dry it and get the circulation going. If the ewe is still uncooperative when the lamb is standing, you can nudge the lamb gently towards the ewe's udder, take a nipple between thumb and forefinger and squeeze, in order to get the milk flowing. If milk comes out, aim the stream right at the lamb's mouth. There are a lot of "ifs" here. If the mother is doing the job, stand back and don't interfere. But the lamb must have colostrum (first milk). Colostrum contains antibodies that help the newborn lamb resist infection and disease and help it get off to a good start.

If the ewe is in labor for more than one hour without bearing a lamb, call your sheep farmer friend or a vet, for advice and help. Most keepers of sheep will go out of their way to help a fellow shepherd.

Now it's time to carry the lamb to the small pen. Walk slowly and the mother will follow. Make sure the floor of the pen is well bedded down with clean straw or hay. Secure them in this snug, draft-free pen, give the ewe a bucket of lukewarm water, and if the lamb seems sturdy and is nursing on its own, leave them alone.

Check on the pair an hour or two later and if all seems well, you've got it made.

Ewes and lambs are usually kept in this small pen for two or three days. The close confinement helps them get acquainted, and if there are complications, it's easier to get ahold of them for treatment. Besides water, you can give the ewe hay the first day or two. After that, begin feeding grain again. If the ewe has a small udder and an extra large lamb, you can give up to 1½ pounds of grain daily.

Daily Care and Observation

Keep feeding hay and grain. Supply fresh water daily. Keep salt available at all times.

Within a couple of weeks, the lamb(s) will begin to nibble at both hay and grain. By the time lambs are a couple of months old, they can eat about ½ pound of grain a day. If both ewes and lambs are fed from the same trough, the lambs won't have a chance at the grain.* You can build a simple creep feeder that will allow the lamb to pass through a small opening to get to the grain, but keeps the adult sheep out. With the combination of milk and grain, the lamb will gain in size and weight very quickly.

Most sheepmen dock the lambs' tails. The long tail of the lamb can gather a lot of burrs and manure in a few months' time. The best way to dock the lamb's tail is to use a device called a band castrator. This instrument is used to place a small rubber ring around the tail, a couple inches below the base. The ring cuts off circulation and within a month, the long tail drops off. This instrument is also used for castration, thus its name, but it is not necessary to castrate your male lambs if you are going to slaughter them within five months. Ram (male) lambs grow faster than ewe lambs or wethers (castrated males), and their carcasses are leaner. As a band castrator costs about $25, try to borrow one from your sheep farmer friend, rather than buy one for just a couple of lambs.

A band castrator may be used to place a rubber band around a lamb's tail. The ring cuts off circulation and the tail drops off.

It is important to check the condition of the ewe's feet during the winter. To do this, catch and hold the sheep immobile and examine each hoof. Trim all excess growth from the sides and front with the hoof knife. Again, consult your shepherd friend.

Trim excess growth from the sides and front of the feet with a knife. The pair at right are trimmed correctly.

---

* See Appendix D for plans for building creep feeder.

If a sheep limps, it may have foot rot. Examine the foot carefully. If the area between the claws is swollen slightly, is overly moist and has an unpleasant odor, foot rot can be suspected. The foot should be trimmed, and one method of treatment is to spray the affected area with formaldehyde. Most livestock supply houses carry formaldehyde in small cans.

Internal Parasites

There are at least 25 different types of worms that can affect sheep. Worms can cause the sheep to be very thin or anemic. They can cause the death of older ewes and young lambs. You should worm your ewes several weeks before they lamb and again before they go out on pasture. There are many good worming agents available at supply stores. Some are in powder form that you mix with water. With this type of mixture, you would hold the sheep's mouth open and pour the required amount of solution down its throat. Follow the directions carefully.

External Parasites

The tick is the most common external parasite that affects sheep. It is not really a tick at all, but a ked or wingless fly. The keds irritate the sheep's skin by sucking its blood. Sheep with ticks (keds) are always rubbing up against posts and walls in order to relieve the itch. Sometimes they will bite at their wool and spoil a portion of it. Sheep with heavy tick infestations appear unthrifty and can lose weight. An inexpensive treatment for ticks is to put the tick-control chemical solution in a common garden sprinkler can and pour it over the sheep. It is easier to treat sheep for ticks after they are sheared and before they go out on pasture.

Shearing

We used to have two big spring storms that occurred with some regularity when I lived in the north country. The first was called the "Robin Storm" and fell upon us in April, the day after all the male robins arrived to stake out their territory for their mates. The second storm was called the "Sheep Storm" and arrived in May, the day after the sheep were sheared. Woe betide those shepherds who turned their sheep out to pasture immediately after shearing to let those skinny, naked bodies shiver in the freezing rain or under a cold blanket of snow. Be that as it may, your ewes have to be sheared, and the best time to do it is before they go on pasture. As you only

The best time to shear is before the sheep go out to pasture.

have two adult sheep, a professional shearer will probably not come to your place. You will have to take them to the closest sheep farm when the shearer is there. Shearing will cost you at least $2 per sheep.

Some breeds of sheep, like the Suffolk, will yield about 5 pounds of fleece. Others, like the Corriedale, have a wool clip that averages from 10 to 12 pounds. You can sell the wool on the market, or, if you don't like the current price, wrap the fleece and store it in a paper-lined bag until the price goes up. You can also use it yourself or sell to local wool spinners.

Pasture

When the new grass in your pasture has reached a height of 3 or 4 inches, you can turn your sheep and lambs out. Before you do, walk your fences and make sure they are in good condition. If there is no stream or creek accessible, carry your water tubs out and fill them. Put the salt block out and secure it on a piece of pipe or stake so that it won't leach away too fast, as it tends to do when placed on the ground.

To prevent digestive upset, give your sheep and lambs a good feed of dry hay before they hit the pasture. About this time, the ewe's milk supply will have peaked, but the lambs will still nurse.

If possible, it would be a good idea to divide your pasture in half, with a cross fence. Then, when the sheep have eaten the grass down in one lot, you can rotate pastures and shift them to the other one. Always provide plenty of fresh water for the flock.

By late July or early August, the lambs should weigh about 100 pounds. Catch a lamb and with a portable scale, estimate its weight. If it's only 70 pounds, let it graze for a few weeks longer. If it's close to 100 pounds, it's time for the animal to provide meat for your table. Don't make pets of the lambs. It's hard to eat a pet.

Slaughtering Lambs

The day before the animal is to be killed, confine it in a clean pen, provide plenty of water, but no feed. Restricting feed results in an easier job of evisceration. Water helps in bleeding out the animal.

The animal is stunned with a quick blow to the head, or shot with a rifle. It is then bled out by cutting of blood vessels. After that, it is hung up and the pelt (wool and skin) is removed by a process called fisting. This consists of literally pulling the pelt off of the lamb, much as you would skin a rabbit. The sheep is then eviscerated and the paunch and intestines removed, along with the heart, liver and kidneys. The carcass should then be chilled for 24 to 48 hours. In my locality, it costs from $10 to $15 to have a lamb slaughtered. The cutting of the carcass into roasts, chops, legs, flank, breast and neck, and the wrapping of these cuts in reasonable portions, costs $.15 per pound.

Dressing Percentage

On the average, the lamb will dress out to about 45 percent, bone and internal organs included. That 100-pound lamb will provide you with about 35 pounds of edible boneless meat.

Manure

Sheep manure is higher in quality and more valuable than either horse or cow manure. It does not burn and can be applied directly to lawns or gardens. Of course, it would make an excellent addition to a compost pile.

Note

If you decide to continue raising sheep, you will need a ram (male sheep) to breed your ewes the following fall. One way to do this is to buy a yearling ram, breed your ewes, and then sell him. This will help recoup the expense.

References and Sources

Animal Science, M. E. Ensminger, The Interstate Publishing Co., Danville, Illinois, 1969. This book, of over 1,200 pages, covers the science of raising domestic farm animals from chickens to horses; there's an excellent section devoted to sheep production.

Raising Sheep the Modern Way, Paula Simmons, Garden Way Publishing Co., Charlotte, Vermont, 1976. A very informative book on raising sheep. It should be on the must list for every beginning shepherd; begins by telling you how to buy sheep and ends with a working calendar of things to do, month by month.

Lamb: Slaughtering, Cutting, Preserving and Cooking on the Farm, Farmer's Bulletin No. 2264, United States Department of Agriculture, Washington, D.C., 1977. This 40-page booklet gives detailed, step-by-step information on how to slaughter, dress out and cut up your own lamb.

Shepherd, Sheffield, Massachusetts. This monthly magazine is a must for the small-flock owner; covers breeding, lambing, feeding and treatment of diseases; the classified ads are a good source for buying sheep and selling wool.

CHAPTER 8

CALVES

If you like beef, and a lot of us do, witness the fact that the per capita consumption of beef in the United States is about 125 pounds a year, you may have considered raising a calf to supply your own meat. Before you plunge into the business of raising a calf, keep in mind it is not a quick venture.

How long does it take? This depends on the size of the animal and how you feed him. If you buy a calf in the fall, a recently weaned animal from six to eight months old, weighing from 400 to 500 pounds, and raise him to slaughter weight by feeding mostly grass and hay, it will take about a year. If you buy a big, but thin, yearling in the spring, one weighing from 600 to 700 pounds, it can take about six months.

In the following pages, I'll use a variety of terms to describe the age, sex and size of cattle. To clarify matters, here are some general definitions:

A *calf* is a young animal less than one year old. A *yearling* is more than one, but less than two years old. *Two-year-olds* are from 24 to 36 months of age.

Besides being classified by age, cattle are also divided into classes according to their sex. A *bull* is an entire male of any age. A *steer* is a castrated male. A *heifer* is a female that has not had a calf. A *cow* is a female that has had at least one calf.

There are also terms that denote the various stages of growth and ultimate purpose of cattle. A *feeder* is a male or female calf, of good size and well-fleshed out (flesh means muscle), ready to be heavily fed in order to complete the process known as finishing. *Finishing* means the putting on of fat. A layer of fat of some proportion is necessary in that it makes beef more tender. *Prime beef* has a moderate amount of fat marbled throughout the meat. When we eat meat, we are eating muscle, but it's the fat that gives it a distinctive taste. The presence of fat on a carcass also aids in our being able to store meat for a longer period of time. *Veal,* the meat from very young calves, has little or no fat and does not keep well.

The term *stocker* is sometimes used to denote a young animal, male or female, that is being held for growth in size and weight, rather than for immediate finishing.

The term *good slaughter weight* means that the animal is ready for butchering. For steers, 1,000 pounds is considered good slaughter weight. Heifers mature faster than steers, and if they are held too long, they tend to put on too much fat. Up to 900 pounds is considered good slaughter weight for heifers.

By the way, you will *not* get 1,000 pounds of steaks, roasts and hamburger from the animal. Cattle dress out to an average of about 60 percent. If the animal is in top condition, you can realize about 465 pounds of meat, some bone included, after it is cut up.

How Much Land Is Necessary?

That all depends upon the quality of the grass (pasture) you have available. Pastures are usually divided into two classes: legumes and grasses. *Legumes* are plants that have the ability to use nitrogen from the atmosphere. Alfalfa, trefoil, vetch, lespedeza and the clovers are all legumes. These plants are higher in protein than most grasses. One acre of legumes would be enough to support one growing beef animal. However, when a legume is used alone as a pasture crop, there is danger of bloat. The animal's stomach becomes distended and death can follow quickly. In order to avoid bloat, there is usually a combination of legumes and grass grown together on a well-tended pasture. As a beginning stockman, it is doubtful that there will be any legume pasture present within the acreage that came along with the place you bought in the country. The cost of the machinery and equipment necessary for planting legumes, or even hiring the work done, would be prohibitive.

The word *grass* covers a wide range of plants, everything from creeping bent to quackgrass, from Kentucky-Blue to bamboo, with oats, rice and rye in between. Plus several hundred species of weeds.

If you live in a semi-arid or desert area, it could take from 5 to 15 acres just to support one beef animal.

If you live in a typical area, and your fields have not been improved in recent years, your pasture will probably contain a mixture of rye grass, wild oats, white (Dutch) clover, milkweed, dandelions, daisies and thistles. If the ground is sour, there will be lots of mossy areas. And, of course, the brambles, bushes, brush, gray birch and poplar (popple) will be encroaching upon the territory like good soldiers, ready to take over.

Thus, the idea of resting the pasture for a couple of years and letting it restore itself does not work. Nature does not permit a vacuum. If *you* don't nurture the pasture, the weeds will gladly spread their seed. I would not recommend the keeping of livestock in the hope of clearing land that is already highly overgrown with brush and small trees. It's a lost cause. But, if the unwanted invaders haven't yet taken over, cattle can keep a pasture close cropped and viable. But, not forever; sooner or later you will have to improve the land with seed and fertilizer.

If your pasture is just average and fairly unimproved, it will take about two acres to support that calf all summer long.

A small, very small, plot of softwoods like spruce, pine, hemlock and other evergreens, enclosed within your pasture, would be an ideal shelter for the calf. It would protect him from the wind and cold rains and you wouldn't need any other shelter.

Dairy cattle have a thin coat of hair and a large amount of udder vulnerable to extremely cold weather. They must be housed in winter. Beef cattle have a thick hide and coat of hair and can survive temperatures of $-70°$ F., as they do with regularity in the Rocky Mountain states.

As you will have to have a place to put the animal when you bring him home, let's discuss fencing.

One way to prod a calf.

Fencing

Keep in mind: the more square the area you fence, the less wire you will need. The distance around one square acre is about 836 feet. The perimeter of two square acres is about 1,200 feet. You can double the acreage fenced with just one-third more wire. I would recommend fencing your two acres with at least three strands of barbed wire. This wire comes in 80-rod rolls (1,320 feet). You would need three rolls to enclose two, fairly square acres. A roll of four-point barbed wire costs about $36. Spacing them 12 feet apart, you would need 100 posts to hold the barbed wire. Untreated, soft-wood posts, 6 feet long and with points, cost about $1 apiece in my area. If you can cut them yourself, you can save a lot of money. Locust and cedar posts last longer and cost more. Steel posts cost over $3 in my area. An un-treated softwood post might last five years. Treated posts cost more and can last 10 years. The fence should be at least 4 feet high. If the animal is of good size and your wire is snug and tight, the three strands should hold him in. He can jump over 4 feet, but unless the grass is a lot greener on the other side, there's no incentive. Four strands of wire would be even better and five would be ideal. It depends on how much you want to spend. If you go with three strands, it would be a good idea to use the leftover wire to make a cross fence and cut the pasture in half. This way you could rotate the calf between the two fields. When the grass got down below 3 inches in one pasture, you would move him to the other one, where the weeds would be getting out of hand. Don't forget to leave space for a gate.

Electric Fence. Instead of barbed wire, you could use a single strand of electric fence. This is a smooth wire that carries an electrical charge in an on-off system, and gives a jolt to anything that touches it. One roll, ¼ mile long and costing about $20, would enclose two acres with something to spare. It can be powered by a 6-volt storage battery that costs about $15. Rather than requiring 100 posts, you can space them wider and perhaps get by with just 60 posts. The wire should be about 30 inches above the ground. One advantage of using electric fence is that it is fairly mobile. If you want to enclose a different pasture, there is only one strand and fewer posts to move, in comparison to a barbed-wire fence. If you use electric fencing, I might suggest stringing one strand of barbed wire, from 8 to 12 inches above the smooth electric wire, for security reasons. When first put out on a pasture, a calf will usually walk the boundaries, looking for a place to get out. It would be clever to introduce him to the fence the first day by holding a handful of clover next to the wire as an offering, and when he leaned against it and got a mild shock, he would give it a wide berth forevermore. One disadvantage of electric fencing is that if brush or dry weeds grow tall underneath it, they can ground it out and render it ineffective. Then the calf is free as a bird. The wire has to be insulated from the posts by small plastic insulators which cost about a dime each.

If there is a creek or stream reasonably close to the pasture, try to enclose a portion of it within your fence. It will save you the trouble of hauling a lot of water in the summertime. On a hot day, the calf will drink several gallons of water. And water is the most important element in aiding natural body processes.

Buying the Calf

I would suggest that you buy the calf from a local farmer, rather than at a livestock auction sale. This way, the chance of it being exposed to diseased animals will be much less. Unless the farmer has a scale, you will have to take his word on the calf's weight. A better idea is to purchase a beef weight tape, costing about $2, from a livestock supply store. This tape is made of non-stretch cloth, marked off in inches, with a corresponding chart that gives an approximate weight of the animal. To use it, place the tape around the heart girth, directly behind the front legs and draw it up snugly, preferably when the calf's head is up. Allowing for a 5 percent error, this tape provides a good estimate of weight.

If you live near a large dairy farm, it is possible to obtain a bob-calf at a very low cost. A bob-calf, so called because of its short tail, is the newborn of a dairy cow. The calf would be two or three days old, and separated from its mother, after having received that most important first milk (colostrum) for a short period. If it's a bull calf, the farmer can't use it because most dairy cows are bred by artificial insemination. With artificial insemination, the dairyman has his choice of the best herd sires available, for a low fee, and doesn't have to care for a sometimes troublesome bull all year around. If it's a heifer, the farmer won't keep it unless it is of special pedigree or breeding or shows exceptional promise. The bob-calf, then, is a by-product. It exists because in order to freshen (give milk) a cow has to be bred and have a calf. The day it's born, the bob-calf has served its purpose.

I would *not* suggest that you buy a bob-calf. It may not turn out to be a bargain. Very young calves are susceptible to many diseases and infections, and there is a lot of stress caused by being separated from the mother. It has been estimated that from 15 to 20 percent of all dairy calves die before reaching maturity. As a beginner, you would not have sufficient know how or experience in the care and feeding of a young dairy calf.

I would suggest, instead, that you buy a calf of one of the medium-sized beef breeds, preferably one of the British breeds. These include the Hereford, which has a white face and red or sometimes reddish-brown body; the Black Angus, which is solid black; or a Shorthorn, which can be all red, all white or red-roan. These British breeds mature early, are of medium size, and require less feed to obtain the desirable slaughter weight that you are aiming for. Don't buy a purebred as they cost more, and you are going to eat the calf, not breed it. Buy a good grade or straight-bred animal.

Another good choice would be to purchase a crossbred, like a Hereford-Angus hybrid, or Angus-Shorthorn cross. Hybrids can do very well, combining the best traits of both breeds. Or you could buy a beef-dairy cross such as a Hereford-Holstein or Angus-Holstein. You could even buy a straight-bred Holstein (dairy breed) steer calf, provided it was of good size, weaned from its mother and accustomed to eating grass and hay. The Holstein could gain well, but would take longer to finish and when dressed out would yield less choice meat, as the conformation of dairy cattle differs from that of beef-type cattle.

If you live in the deep south or southwest, a calf with Brahman blood would be an excellent choice. The Brahman originated in India, but a definite American breed has been developed over the years. The Brahman is usually red, brown or gray, although there are many with various spots and other colorations. The Brahman has long, droopy ears, much pendulous skin under the throat and a pronounced hump over the shoulders. Contrary to the myth of Brahman bulls being wild and untameable (a myth perpetrated by rodeo promoters), the Brahman can be as gentle as any other breed. The Brahman or a Brahman-cross calf would be more resistant to ticks and the disease called *pinkeye*, which can have a disastrous effect on cattle of the British breeds.

Some of the calves available would have names like Braford (Brahman-Hereford cross) or Brangus (Brahman-Angus cross).

Hereford calves.

Whatever you buy, make sure that the calf is sound and moves well. He should be alert, with bright eyes. He should be long, with plenty of leg under him, and the legs should be placed well apart. The hindquarter of a beef steer produces a large percentage of the choicest meat. His rear end should be thick, round and well muscled.

Whether you buy a male or female doesn't make too much difference, although the heifer will finish quicker, but at a lighter weight. If you buy a male, he should be castrated, thus, a steer. Bulls grow faster and the carcass is leaner, but as they grow bigger and older, they have a tendency to go through fences looking for heifers to breed.

If you buy in the fall, the calf will be about six months old, weigh from 400 to 500 pounds, be weaned from its mother and accustomed to eating forage. The calf will cost less in the fall, as more cattle go to market at that time and the supply is greater. Last fall, in my area, calves of the above description were available for about $.55 per pound. Thus, a 450-pound calf would cost $247.50.

The previous spring, a big, but thin calf approaching yearling age and weighing about 650 pounds, cost about $.60 per pound, or $390. This yearling would have been roughed through the winter on hay alone and be ready to go out on pasture and gain weight on the new grass. In the spring, feeders or stockers are less plentiful and cost more.

If you buy a calf in the fall, you will have to provide some kind of shelter for it, nothing fancy, just a place to get out of the wind or a blizzard. You will also have to feed it about 60 bales of hay to carry it through the winter until it can go out on pasture. Supply plenty of water and provide a block of salt. At an average of $1.50 per bale, the hay will cost $90. The salt sells for about $4.

When you do buy a calf, try to have the cost of delivery included in the purchase price. This will save you the cost of hiring a truck to transport the animal to your place.

Care and Observation

Whether you buy in spring or fall, calves will spend the summer out on pasture. If there is no natural source of water, you will have to provide a water trough or tub and keep it filled. On a hot day, a calf will drink 8 gallons of water. Also, calves need a block of salt. And the calf must have protection from the hot sun, whether it's in the form of a small grove of trees or a simple arrangement of boards nailed on top of upright posts.

Check your fences regularly, repairing if necessary.

Calves are the hardiest of young farm animals. If the animal appears to have a physical problem, contact a neighboring cattleman or a vet for advice and treatment. After all, you have a sizable investment in the animal.

If you divide your pasture in two, rotate the calf between the two fields before the grass gets down too low in one and too high in the other.

Depending on the quality of the grass in your pasture, the calf will harvest from 1½ to 2 tons of forage over the summer. The 650-pound yearling purchased in the spring will come into the fall weighing about 850 pounds. Use your beef weight tape to estimate his weight. The best time to do this is at feeding time.

If you bought in the previous fall, the calf would weigh about the same, but remember, you had to carry him through the winter. You could butcher the animal at this weight, but he would be very lean and not as juicy in the oven. When the grass is gone and he comes off pasture in September or October, switch him to a finishing ration, with a small amount of hay, to put some finish on the steer or heifer. Make the feed changes very gradually, and remember, with any animal, it is better to underfeed than to overfeed.

Finishing Rations

There is a wide variety of supplemental feeds that can be used for finishing cattle. Corn silage, cull potatoes, chopped turnips, beet pulp, apple pomace, sugar cane and spent brewers' mash are all good. If you don't have access to these products, use grain. Assuming that you don't have the equipment for growing and harvesting grain, or the land, you will have to buy it.

It will take at least 600 pounds of grain, plus 20 bales of good hay, fed over a period of two months, to bring the animal up to a good slaughter weight, with a decent finish. At current prices, the grain will cost about $60 and the hay $30.

What It Will All Cost

There are several variables here. If you are able to use electric fencing, you can cut the cost considerably. If there is a wider spread in the difference of the price of a calf purchased in the fall, in relation to the cost of a calf bought in the spring, it can change the figures drastically.

As an example, using my figures, the following would cover the cost from start to finishing if electric fencing is used and the calf is purchased in the fall:

450-lb. calf purchased in fall	$247.50
1 roll electric fence wire	20.00
1 roll barbed wire	36.00
60 posts	60.00
60 insulators	6.00
1 15-volt battery	15.00
1 weight tape	2.00
60 bales of hay for wintering	90.00
1 block stock salt	4.00
600 lbs. grain for finishing calf	60.00
20 bales of hay for finishing calf	30.00
	$570.50

The calf purchased in the spring would cost about $50 more to raise, using these figures, but, again, it all depends on the price differential between the two calves. One important thing, the older calf would not have to be fed and watered every day, all winter long. As soon as you bought him, he would go out on pasture.

Slaughtering

When you wrap the weight tape around the steer's girth and it reads 73 or 74 inches, he should weigh about 1,000 pounds. It's time to make arrangements for your local slaughterhouse or butcher, to do the job.

The animal is stunned with a blow to the head or shot between the eyes with a rifle. It is then hoisted up, stuck with a knife and bled out. The head is removed and the hide skinning process begun. An incision is made down the belly, from rib cage to rear end, and all internal organs of the body are removed. The carcass is then split in half and washed down. It is then put in a cooler for at least 24 hours before any further cutting is done. Chilling makes the carcass firm and easier to cut up.

The 1,000-pound steer will dress out to about 600 pounds. From that 600 pounds, it will yield about 465 pounds of usable meat, some bone included. You will also get about 16 pounds of liver, a 5-pound heart, 4 pounds of kidneys, about 3 pounds of tongue, a pound of sweetbreads and a hide weighing about 70 pounds, which you can tan or sell.

In my area, it costs $15 to have a steer slaughtered and $.15 a pound for cutting and wrapping.

The following is a guide showing the breakdown of retail cuts yielded by 465 pounds of meat and their value at current meat market prices:

Retail Cut	Pounds	Price	Total
Porterhouse, T-bone and club steaks	35	$3.99	$139.65
Sirloin steak	40	2.99	119.60
Round steak	65	2.99	194.35
Rib roast	45	2.59	116.55
Boneless rump roast	25	2.99	74.75
Chuck roast	100	2.19	219.00
Hamburger	45	1.59	71.55
Stew beef and miscellaneous cuts	110	1.79	196.90
	465		$1,132.35
Bone, fat and waste	135		
	600 lbs.		

Your steer, then, is worth $1,132.35 at retail meat market prices. If your start-up to finishing costs, as shown on the previous page, amount to

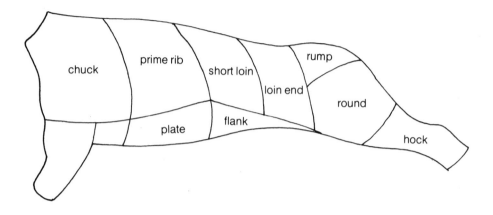

$570.50 and you add a slaughtering fee of $15, for a total of $585.50, that's not a bad deal.

Retail value	$1,132.35
Your cost	− 585.50
	$ 546.85

Of course, this is only an example and things seldom work out in reality as they do on paper. But it will give you a general idea of what it can cost and what you can save by raising your own calf for beef. It does not take into account your time and labor to feed and water the calf all winter long. Or the cost of electric power to run your well pump. Or all the work you did in fencing. Speaking of fencing, if you decide to raise another calf, your fence is already up. You've got the posts and wire, all you have to do is buy another battery. You will save $122 the next time around, barring a little fence repair here and there.

Manure

If you wintered the calf, manure collected from his shelter would make fine fertilizer for your vegetable garden the next spring. On pasture, it is fairly well distributed by the animal.

References and Sources

Animal Science, M. E. Ensminger, The Interstate Publishing Co., Danville, Illinois, 1969. This book covers the science of raising domestic farm animals; there are several fine chapters devoted to beef-cattle production.

Raising a Calf for Beef, Phillis Hobson, Garden Way Publishing Co., Charlotte, Vermont, 1976. This book covers all the basics of raising a beef calf, including housing, feeds, medical care and butchering.

GENERAL COMPARISON OF

Species	Hardiness	Basic unit	Estimated cost	Estimated feed consumption	Estimated feed cost
Broiler chickens	Fair	25 day-old chicks	$20	200 lbs.	$22
Turkeys	Poor	6 day-old poults	$18	300 lbs.	$35
Ducks	Good	15 day-old ducklings	$20	280 lbs.	$37
Geese	Excellent	8 day-old goslings	$32	300 lbs.	$38
Rabbits	Fair to good	3 adults	$45	128 lbs. for 8 young	$19.20
Pigs	Good	1 weanling	$35	650 lbs.	$60
Lambs	Poor to fair	2 bred females	$150	Milk & grass (2 lambs)	$60 for 2 adults
Calves	Excellent	1 calf, 6 months old	$247.50	600 lbs. grain, 80 bales hay (pastured in summer)	$180

POULTRY AND ANIMALS

Time involved	Estimated live weight at butchering	Average dressing %	Estimated edible meat and some bone included
8 weeks	4 lbs. per chicken	75%	75 lbs.
5 months	18 lbs. per turkey	78%	75 lbs.
7–8 weeks	7 lbs. per duck	70%	75 lbs.
12–14 weeks	12 lbs. per goose	70%	68 lbs.
8 weeks	4.5 lbs. per rabbit	55%	20 lbs.
5 months	200 lbs. per pig	75%	150 lbs.
5 months	80–100 lbs. per lamb	45%	70 lbs. (two lambs)
12 months	1,000 lbs. per calf	60%	465 lbs.

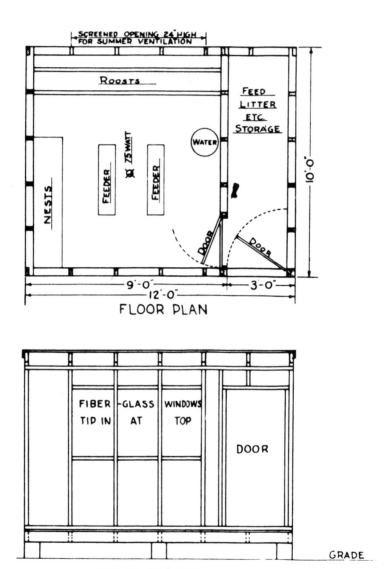

FLOOR PLAN

FRONT FRAMING

SCALE

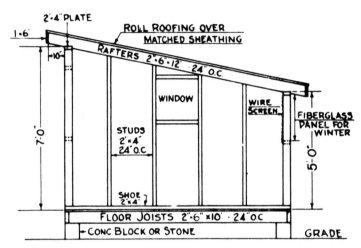

END FRAMING

Labels within the figure:

2"-4" PLATE
1-6'
ROLL ROOFING OVER MATCHED SHEATHING
-10"
RAFTERS 2"-6"-12' 24" O.C.
WINDOW
WIRE SCREEN
FIBERGLASS PANEL FOR WINTER
7'-0"
STUDS 2"×4" 24" O.C.
5'-0"
SHOE 2"×4"
FLOOR JOISTS 2"-6"×10' 24" O.C.
CONC BLOCK OR STONE
GRADE

BILL OF MATERIALS

FOUNDATION		6 CONCRETE BLOCKS 8"-8"-16"
FLOOR JOISTS		7 PC 2"-6"×10'
FR & REAR SILLS		2 PC 2"×6"×12'
FLOOR		150 BD FT T&G SHEATHING
SHOE		54 LIN FT 2"×4"
STUDS	REAR	9 PC 2"×4"×5'
	FRONT	9 PC 2"×4"×7'
	ENDS	4 PC 2"×4"×12'
	PARTITION	2 PC 2"×4"×12'
PLATES		2 PC 2"×4"×12'
ROOF		175 BD FT T&G SHEATHING
		1½ SQ ROLL ROOFING
SIDING & DOORS		450 BD FT T&G SHEATHING
WINDOWS		2 PC 2'×10' FIBERGLASS (FLAT)
MISCL. FRAMING		4 PC 2×4×12'
NAILS & HDWRE		

Cooperative Extension Service, Agricultural Engineering Department, University of New Hampshire, United States Department of Agriculture Cooperating.

Appendix C

Addresses of Extension Service Offices by State

ALABAMA

Auburn University, Auburn, Alabama 36830
102 Federal Bldg. P.O. Box 15, Cullman, Alabama 35055

ARIZONA

Agricultural Science Building, Univ. of Arizona, Tucson, Arizona 85721

ARKANSAS

University of Arkansas, P.O. Box 391, Little Rock, Arkansas 72203
University of Arkansas, Dept. Animal Sci. R-C123, Fayetteville, Arkansas 72701

CALIFORNIA

University of California, Department of Avian Sciences, Davis, California 95616
University of California, 306 Agricultural Extension Building, Riverside, California 92502
P.O. Box 1411, Modesto, California 95353
Agr. Res. & Ext. Center, 9240 S. Riverbend Ave., Parlier, California 93648
Suite 202, 21160 Box Springs Rd., Riverside, California 92507
566 Lugo Avenue, San Bernardino, California 92415
Building 4, 5555 Overland Avenue, San Diego, California 92123
Room 100-P, Mendocino Avenue, Santa Rosa, California 95401
684 Buena Vista St., Ventura, California 93001
1000 S. Harbor Blvd., Anaheim, California 92805

COLORADO

Colorado State University, Fort Collins, Colorado 80521

CONNECTICUT

University of Connecticut, Storrs, Connecticut 06268
Agricultural Center, Haddam, Connecticut 06492
24 Hyde Avenue Rt. 30, Rockville, Connecticut 06066
562 New London Turnpike, Norwich, Connecticut 06360

DELAWARE

University of Delaware, RD 2, Box 48, Georgetown, Delaware 19947
Agr. Hall, Newark, Delaware 19711

FLORIDA

University of Florida, Gainesville, Florida 32603
Chipley, Florida 32428

GEORGIA

University of Georgia, Athens, Georgia 30602
Calhoun, Georgia 30701
Oakwood, Georgia 30566
Tifton, Georgia 31794

HAWAII

University of Hawaii, 1825 Edmondson Rd., Honolulu, Hawaii 96822

IDAHO

University of Idaho, Moscow, Idaho 83843

ILLINOIS

University of Illinois, 322 Mumford Hall, Urbana, Illinois 61801

INDIANA

Purdue University, Lafayette, Indiana 47907

IOWA

Iowa State University, Kildee Hall, Ames, Iowa 50010

KANSAS

Kansas State University, Leland Call Hall, Manhattan, Kansas 66506

KENTUCKY

University of Kentucky, Lexington, Kentucky 40506
Somerset Community Col., 808 Monticello Rd., Somerset, Kentucky
42501
1270 Montgomery Avenue, Ashland, Kentucky 41101

LOUISIANA

Louisiana State University, Knapp Hall, Baton Rouge, Louisiana 70803

MAINE

University of Maine, Hitchner Hall, Orono, Maine 04473
P.O. Building, Belfast, Maine 04915
P.O. Building, Lewiston, Maine 04241
Federal Bldg. Rm. 209, Rockland, Maine 04841

MARYLAND

University of Maryland, Dept. of Poultry Science, College Park, Maryland 20742
Broiler Sub-Station, RFD 5, Salisbury, Maryland 21801

MASSACHUSETTS

University of Mass., 314 Stockbridge Hall, Amherst, Massachusetts 01002

MICHIGAN

Michigan State University, 113 Anthony Hall, East Lansing, Michigan 48823
Coop Extension Service, P.O. Box 79, Zeeland, Michigan 49464

MINNESOTA

University of Minnesota, St. Paul, Minnesota 55101

MISSISSIPPI

Mississippi State University, P.O. Box 5425, Mississippi State, Mississippi 39762
Box 9714, Jackson, Mississippi 39206

MISSOURI

University of Missouri, Columbia, Missouri 65201
RD1 Bldg. Belcrest & E. Trafficway, Springfield, Missouri 65802

NEBRASKA

University of Nebraska, Lincoln, Nebraska 68503

NEW HAMPSHIRE

University of New Hampshire, 55 Pleasant St., Rm. 331, Concord, N.H. 03301
Kendall Hall, Durham, New Hampshire 03824

NEW JERSEY

Rutgers—The State University, CAES P.O. Box 231, New Brunswick, New Jersey 08903

NEW MEXICO

New Mexico State University, Dept. Poultry Sci., Box 3P, Las Cruces, New Mexico 88001

NEW YORK

Cornell University, Rice Hall, Ithaca, New York 14850
249 Highland Ave., Rochester, New York 14620
380 Federal Building, Syracuse, New York 13202
Coop Ext. Regional Office, Martin Rd., Voorheesville, N.Y. 12186

NORTH CAROLINA

North Carolina State University, P.O. Box 5307 Scott Hall, Raleigh, N.C. 27607

NORTH DAKOTA

North Dakota State Univ. Stevens Hall, Rm. 218A, Fargo, North Dakota 58102

OHIO

Ohio State Univ., 2120 Fyffe Road, Columbus, Ohio 43210

OKLAHOMA

Oklahoma State University, Stillwater, Oklahoma 74074

OREGON

Oregon State University, Poultry Sci. Dept., Corvallis, Oregon 97331

PENNSLYVANIA

Pennsylvania State University, University Park, Pennsylvania 16802

PUERTO RICO

University of Puerto Rico, Rio Piedras, Puerto Rico 00928

RHODE ISLAND

University of Rhode Island, Kingston, Rhode Island 02881

SOUTH CAROLINA

Clemson University, Clemson, South Carolina 29631
P.O. Box 378, York, South Carolina 29745
P.O. Box 1711, Columbia, South Carolina 29201

SOUTH DAKOTA

South Dakota State University, Brookings, South Dakota 57006

TENNESSEE

University of Tennessee, P.O. Box 1071, Knoxville, Tennessee 37901

TEXAS

Texas A & M University, College Station, Texas 77843

Res. & Ext. Center, Box 220, Overton, Texas 75684

UTAH

Utah State University, Logan, Utah 84322

VERMONT

University of Vermont, Burlington, Vermont 05401

VIRGINIA

Virginia Polytechnic Institute, Blacksburg, Virginia 24061

WASHINGTON

Washington State University, Pullman, Washington 99163

Western Washington, Res. & Ext. Center, Puyallup, Washington 98371

WEST VIRGINIA

West Virginia University, Morgantown, West Virginia 26506

WISCONSIN

University of Wisconsin, Dept. Poultry Sci., Col. of Agr. & Life Sci., 1675 Observatory Dr., Madison, Wisconsin 53706

WYOMING

University of Wyoming, Univ. Station, P.O. Box 3354, Laramie, Wyoming 82071

WASHINGTON, D.C.

Extension Service—USDA, Room 5509 South Building, Washington, D.C. 20250

Appendix D

Fencing

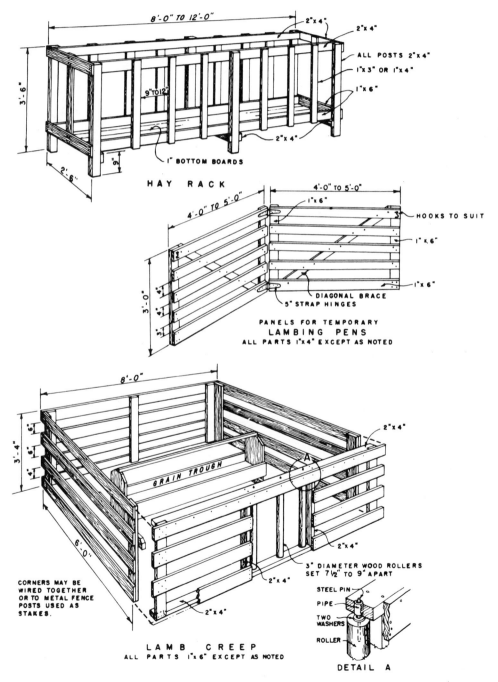

8'-0" TO 12'-0"

3'-6"

2"x4"

2"x4"

ALL POSTS 2"x4"

1"x3" OR 1"x4"

1"x6"

9"TO12"

2"x4"

2'-6"

9"

1" BOTTOM BOARDS

HAY RACK

4'-0" TO 5'-0"

4'-0" TO 5'-0"

1"x6"

HOOKS TO SUIT

1"x6"

3'-0"

4"

4"

3"

DIAGONAL BRACE

5" STRAP HINGES

1"x6"

PANELS FOR TEMPORARY
LAMBING PENS
ALL PARTS 1"x4" EXCEPT AS NOTED

8'-0"

3'-4"

6"

5"

4"

2"x4"

GRAIN TROUGH

A

2"x4"

6'-0"

CORNERS MAY BE
WIRED TOGETHER
OR TO METAL FENCE
POSTS USED AS
STAKES.

2"x4"

2"x4"

3" DIAMETER WOOD ROLLERS
SET 7½" TO 9" APART

STEEL PIN

PIPE

TWO
WASHERS

ROLLER

2"x4"

LAMB CREEP
ALL PARTS 1"x6" EXCEPT AS NOTED

DETAIL A

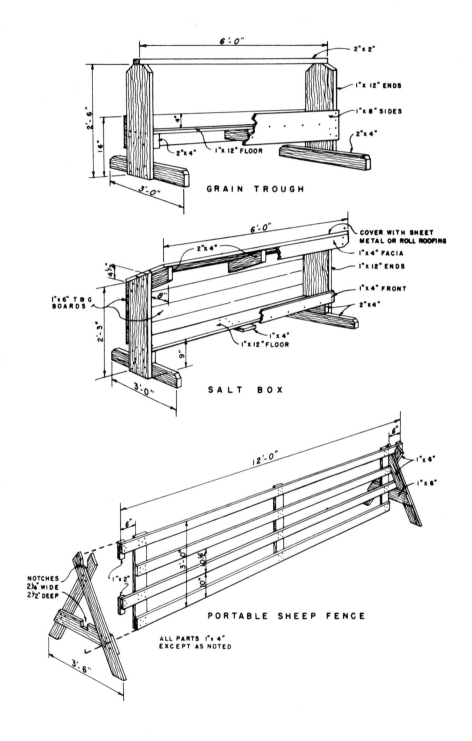

GRAIN TROUGH

SALT BOX

PORTABLE SHEEP FENCE

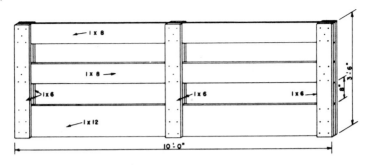

PANEL FOR LAMBS

NOTE: MAY BE USED FOR FENCING A LOT
AND AS A FEED RACK

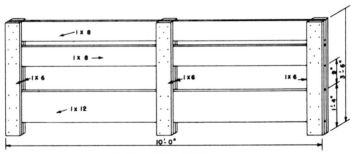

PANEL FOR EWES

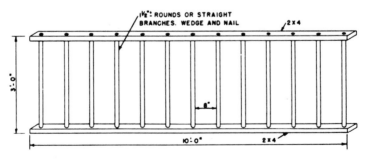

CREEP PANEL

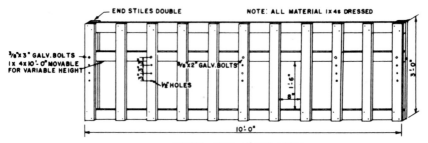

PANEL FOR CREEP FEEDING

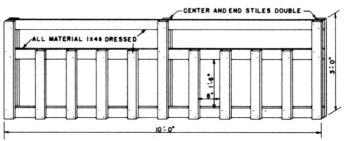

PANEL FOR CREEP FEEDING

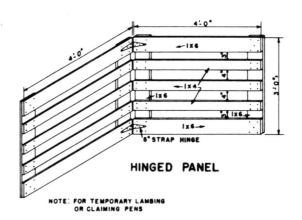

HINGED PANEL

NOTE: FOR TEMPORARY LAMBING
OR CLAIMING PENS

INDEX